THEORY AND REALITY

SCIENCE AND ITS CONCEPTUAL FOUNDATIONS A SERIES EDITED BY DAVID L. HULL

an introduction to the philosophy of science

THEORY AND REALITY

PETER GODFREY-SMITH

The University of Chicago Press / Chicago and London

The University of Chicago Press, Chicago 60637
The University of Chicago Press, Ltd., London

Printed in the United States of America

18 17 14

ISBN-13: 978-0-226-30062-7 (cloth)
ISBN-13: 978-0-226-30063-4 (paper)
ISBN-10: 0-226-30062-5 (cloth)
ISBN-10: 0-226-30063-3 (paper)

Library of Congress Cataloging-in-Publication Data

Godfrey-Smith, Peter.
Theory and reality : an introduction to the philosophy of science / Peter Godfrey-Smith.
p. cm. — (Science and its conceptual foundations)
Includes bibliographical references and index.
ISBN 0-226-30062-5 (alk. paper) — ISBN 0-226-30063-3 (pbk. : alk. paper)
1. Science—Philosophy. I. Title. II. Series.
Q175 .G596 2003
501—dc21
2002155305

⊗ The paper used in this publication meets the minimum requirements of the American National Standard for Information Sciences—Permanence of Paper for Printed Library Materials, ANSI Z39.48-1992.

For my parents

Contents

Preface

This book is based mainly on lectures given at Stanford University during the last eleven years. So the book is a distillation of lectures, but not only of lectures. It also bears the influence of innumerable comments, questions, and papers by students over that time, together with remarks made by colleagues and friends.

The book is written primarily for students, but it is intended to be accessible to a fairly wide audience. I assume no background knowledge in philosophy at all in the reader. My primary aim is to introduce some of the main themes in the philosophy of science, while simultaneously telling an accessible and interesting story about how the field has developed in the last one hundred years or so. In telling this story I have been led to describe the connections between philosophy and other disciplines, and the changing intellectual climate in which theories of science have been offered, in more detail than many introductory books do. I have also tried, in some places, to capture some of the atmosphere of the debates, and the personalities of the protagonists.

Another aim of the book is the outline and defense of a particular point of view, but I have concentrated that discussion mostly in the final third of the book. Philosophy of science seems to me to be still in a state of considerable ferment. That poses a choice for the author of a book like this; one can either abstract away from the disorder and uncertainty, and lay down one particular vision, or one can use the disputes to tell a story about the field—how did we get to where we are now? I have mostly chosen the latter approach. This feature of the book is partly due to the inspiration of John Heilbroner's classic history of economic thought, *The Worldly Philosophers*.

For comments on this work I am very grateful to Fiona Cowie, Michael Devitt, Stephen Downes, Richard Francis, Michael Friedman, Lori Gruen, Tania Lombrozo, Denis Philips, J. D. Trout, Allen Wood, and Rega Wood.

Two anonymous referees for the University of Chicago Press also made helpful criticisms.

For detailed and exceptionally useful comments on entire near-final drafts, resulting in numerous improvements, I am indebted to Karen Bennett, Kim Sterelny, and Michael Weisberg.

Other improvements resulted, as always, from the insight, good sense, deft touch, and unique perspective of David Hull, the editor of the Science and Its Conceptual Foundations series. At the University of Chicago Press, Susan Abrams was constantly enthusiastic about the project and did a great job throughout. It is a rare pleasure to work with an editor like Susan. I am also grateful to Stanford University for much financial and intellectual support over the last eleven years. This support has included several grants, including most recently a Martha Sutton Weeks Fellowship.

Finally, as this is a book written primarily for students, this seems an appropriate place for me to express my enduring gratitude to the four fundamental mentors who taught, guided, and encouraged me when I was a student: Kim Sterelny, Michael Devitt, Stephen Stich, and Philip Kitcher.

A Note for Those Teaching with the Book

The book is organized chronologically, especially until chapter 10, and following the chronology is probably the most appropriate way to teach a course using the book. However, there is also a way to use the book in a course that follows a more thematic organization. Approached this way, chapters 1 and 2 are background; chapters 3, 4, 10, and 14 form a block focused on issues about evidence, testing, and theory choice; chapters 5–11 discuss scientific change and the social organization of science, along with the interaction between these topics and epistemological questions; chapters 12 and 13 address issues more on the metaphysical than the epistemological side of the philosophy of science. The book might also be used, of course, as a supplement to lectures and readings with a very different organization.

The "Further Reading" sections tend to contain quite a lot of primary material, including some difficult works and works intended to give the flavor of recent discussion (such as papers from the Proceedings of the Philosophy of Science Association). This is especially true of the later chapters. The level of difficulty found in the "Further Reading" escalates more quickly than the difficulty of the text itself (or so I hope). The glossary, in contrast, is intended to be very elementary, a tool for those coming to the book with little or no background in the area.

1

Introduction

1.1 Setting Out

This book is a survey of roughly one hundred years of argument about the nature of science. We'll look at a hundred years of argument about what science is, how it works, and what makes science different from other ways of investigating the world. Most of the ideas we will examine fall into the field called "philosophy of science," but we will also spend a good deal of time looking at ideas developed by historians, sociologists, psychologists, and others.

The book mostly has the form of a "grand tour" through the decades; ideas will be discussed in roughly the order in which they appeared. Note the word "roughly" in the previous sentence; there are exceptions to the historical structuring of the book, and I will point out some of them as they arise.

Why is it best to start with older ideas and work through to the present? One reason is that the historical development of general ideas about science is itself an interesting topic. Another reason is that the philosophy of science has been in a state of fermentation and uncertainty in recent years. A good way to understand the maze of options and opinions in the field at the moment is to trace the path that brought us to the state we're in now. But this book does not only aim to introduce the options. I will often take sides as we go along, trying to indicate which developments were probably wrong turns or red herrings. Other ideas will be singled out as being on the right track. Then toward the end of the book, I will start trying to put the pieces together into a picture of how science works.

Philosophy is an attempt to ask and answer some very basic questions about the universe and our place within it. These questions can sometimes seem far removed from practical concerns. But the debates covered in this book are not of that kind. Though these debates are connected to the most abstract questions about thought, knowledge, language, and reality, they have also turned out to have an importance that extends well outside of philosophy. They have made a difference to developments in many other academic fields, and some of the debates have reverberated much further, affecting discussions of education, medicine, and the proper place of science in society.

In fact, throughout the latter part of the twentieth century, all the fields concerned with the nature of science went on something of a roller-coaster ride. Some people thought that work in the history, philosophy, and sociology of science had shown that science does not deserve the dominating role it has acquired in Western cultures. They thought that a set of myths about the trustworthiness and superiority of mainstream science had been thoroughly undermined. Others disagreed, of course, and the resulting debates swirled across the intellectual scene, frequently entering political discussion as well. From time to time, scientific work itself was affected, especially in the social sciences. These debates came to be known as the "Science Wars," a phrase that conveys a sense of how heated things became.

The Science Wars eventually cooled down, but now, as I write these words, it is fair to say that there is still a great deal of disagreement about even the most basic questions concerning the nature and status of scientific knowledge. These disagreements usually do not have much influence on the day-to-day practice of science, but sometimes they do. And they have huge importance for general discussions of human knowledge, cultural change, and our overall place in the universe. This book aims to introduce you to this remarkable series of debates, and to give you an understanding of the present situation.

1.2 The Scope of the Theory

If we want to understand how science works, it seems that the first thing we need to do is work out what exactly we are trying to explain. Where does science begin and end? Which kinds of activity count as "science"?

Unfortunately this is not something we can settle in advance. There is a lot of disagreement about what counts as science, and these disagreements are connected to all the other issues discussed in this book.

There is consensus about some central cases. People often think of physics as the purest example of science. Certainly physics has had a heroic history and a central role in the development of modern science. Molecular biology, however, is probably the science that has developed most rapidly and impressively over the past fifty years or so.

These seem to be central examples of science, though even here we encounter hints of controversy. A few have suggested that theoretical physics is becoming less "scientific" than it used to be, as it is evolving into an esoteric, mathematical model–building exercise that has little contact with the real world (Horgan 1996). And molecular biology has recently been acquiring connections with business and industry that make it, in the eyes of some, a less exemplary science than it once was. Still, examples like these give us a natural starting point. The work done by physicists and molecular biologists when they test hypotheses is science. And playing a game of basketball, no matter how well one plays, is not doing science. But in the area between these clear cases, disagreement reigns.

At one time the classification of economics and psychology as sciences was controversial. Those fields have now settled into a scientific status, at least within the United States and similar countries. (Economics retains an amusing qualifier; it is often called "the dismal science," a phrase due to Thomas Carlyle.) There is still a much-debated border region, however, and at the moment this includes areas like anthropology and archaeology. At Stanford University, where I teach, this kind of debate was one element of a process in which the Department of Anthropology split into two separate departments. Is anthropology, the general study of humankind, a fully scientific discipline that should be closely linked to biology, or is it a more "interpretive" discipline that should be more closely connected to the humanities?

The existence of this gray area should not be surprising, because in contemporary society the word "science" is a loaded and rhetorically powerful one. People will often find it a useful tactic to describe work in a borderline area as "scientific" or as "unscientific." Some will call a field scientific to suggest that it uses rigorous methods and hence delivers results we should trust. Less commonly, but occasionally, a person might call an investigation scientific in order to say something negative about it—to suggest that it is dehumanizing, perhaps. (The term "scientistic" is more often used when a negative impression is to be conveyed.) Because the words "science" and "scientific" have these rhetorical uses, we should not be surprised that people constantly argue back and forth about which kinds of intellectual work count as science.

The history of the term "science" is also relevant here. The current uses of the words "science" and "scientist" developed quite recently. The word "science" is derived from the Latin word "scientia." In the ancient, medieval, and early modern world, "scientia" referred to the results of logical demonstrations that revealed general and necessary truths. Scientia could be gained in various fields, but the kind of proof involved was what we would now mostly associate with mathematics and geometry. Around the seventeenth century, when modern science began its rise, the fields that we would now call science were more usually called "natural philosophy" (physics, astronomy, and other inquiries into the causes of things) or "natural history" (botany, zoology, and other descriptions of the contents of the world). Over time, the term "science" came to be used for work with closer links to observation and experiment, and the association between science and an ideal of conclusive proof receded. The current senses of the term "science" and the associated word "scientist" are products of the nineteenth century.

Given the rhetorical load carried by the word "science," we should not expect to be able to lay down, here in chapter 1, an agreed-on list of what is included in science and what is not. For now we will have to let the gray area remain gray.

A further complication comes from the fact that philosophical (and other) theories differ a lot in how broadly they conceive of science. Some writers use terms like "science" or "scientific" for any work that assesses ideas and solves problems in a way guided by observational evidence. Science is seen as something found in all human cultures, even though the word is a Western invention. But there are also views that construe "science" more narrowly, seeing it as a cultural phenomenon that is localized in space and time. For views of this kind, it was only the Scientific Revolution of the sixteenth and seventeenth centuries in Europe that gave us science in the full sense. Before that, we find the initial "roots" or precursors of science in ancient Greece, some contributions from the Arab world and from the Scholastic tradition in the late Middle Ages, but not much else. So this is a view in which science is treated as a special social institution with a definite history. Science is something that descends from specific people and places, and especially from a key collection of Europeans, including Copernicus, Kepler, Galileo, Descartes, Boyle, and Newton, who all lived in the sixteenth and seventeenth centuries.

To set things up this way is to see science as *unlike* the kinds of investigation and knowledge that routinely go along with farming, architecture, and other kinds of technology. So a view like this need *not* claim that people in nonscientific cultures must be ignorant or stupid; the idea is that in order to understand *science,* we need to distinguish it from other kinds of investigation of the world. And we need to work out how one approach to knowledge developed by a small group of Europeans turned out to have such spectacular consequences for humanity.

As we move from theory to theory in this book, we will find some people construing science broadly, others narrowly, and others in a way that lies in between. But this does not stop us from outlining, in advance, what kind of understanding we would eventually like to have. However we choose to use the word "science," in the end we should try to develop *both*

1. a general understanding of how humans gain knowledge of the world around them *and*
2. an understanding of what makes the work descended from the Scientific Revolution *different* from other kinds of investigation of the world.

We will move back and forth between these two kinds of questions throughout the book.

Before leaving this topic, there is one other possibility that should be mentioned. How confident should we be that all the work we call "science," even in the narrower sense described above, has that much in common? One of the hazards of philosophy is the temptation to come up with theories that are too broad and sweeping. "Theories of science" need to be scrutinized with this problem in mind.

1.3 What Kind of Theory?

This book is an introduction to the philosophy of science. But most of the book focuses on one set of issues in that field. Within the philosophy of science, we can distinguish between *epistemological* issues and *metaphysical* issues (as well as issues that fall into neither category). Epistemology is the side of philosophy that is concerned with questions about knowledge, evidence, and rationality. Metaphysics, a more controversial part of philosophy, deals with general questions about the nature of reality. Philosophy of science overlaps with both of these.

Most of the issues discussed in this book are, broadly speaking, epistemological issues. For example, we will be concerned with questions about how observational evidence can justify a scientific theory. We will also ask whether we have reason to hope that science can succeed in describing the world "as it really is." But we will occasionally encounter metaphysical issues, and issues in the philosophy of language. The discussion will intersect with work in the history of science and other fields as well.

All of philosophy is plagued with discussion and anxiety about how philosophical work should be done and what a philosophical theory should try to do. So we will have to deal with disagreement about the right *form* for a philosophical theory of science, and disagreement about which questions philosophers should be asking. One obvious possibility is that we might try for an understanding of scientific *thinking*. In the twentieth century, many philosophers rejected this idea, insisting that we should seek a *logical* theory of science. That is, we should try to understand the abstract structure of scientific theories and the relationships between theories and evidence. A third option is that we should try to come up with a *methodology*, a set of rules or procedures that scientists do or should follow. In more recent years, philosophers influenced by historical work have wanted to give a general theory of scientific *change*.

A distinction that is very important here is the distinction between *descriptive* and *normative* theories. A descriptive theory is an attempt to describe what actually goes on, or what something is like, without making value judgments. A normative theory does make value judgments; it talks about what should go on, or what things should be like. Some theories about science are supposed to be descriptive only. But most of the views we will look at do have a normative element, either officially or unofficially. When assessing general claims about science, it is a good principle to constantly ask: "Is this claim intended to be descriptive or normative, or both?"

For some people, the crucial question we need to answer about science is whether or not it is "objective." But this term has become an extremely slippery one, used to mean a number of very different things. Sometimes objectivity is taken to mean the absence of bias; objectivity is impartiality or fairness. But the term "objective" is also often used to express claims about whether the *existence* of something is independent of our minds. A person might wonder whether there really is an "objective reality," that is to say, a reality that exists regardless of how people conceptualize or describe it. We might ask whether scientific theories can ever describe a reality that exists in this sense. Questions like that go far beyond any issue about the absence of bias and take us into deep philosophical waters.

Because of these ambiguities, I will often avoid the terms "objective" and "objectivity." But the questions that tend to be asked using those terms will be addressed, using different language, throughout the book. And I will return to "objectivity" in the final chapter.

Another famous phrase is "scientific method." Perhaps this is what most people have in mind when they imagine giving a general theory of science. The idea of describing a special method that scientists do or should follow is old. In the seventeenth century, Francis Bacon and René Descartes, among others, tried to give detailed specifications of how scientists should proceed. Although describing a special scientific method looks like a natural thing to try to do, during the twentieth century many philosophers and others became skeptical about the idea of giving anything like a recipe for science. Science, it was argued, is too creative and unpredictable a process for there to be a recipe that describes it—this is especially true in the case of great scientists like Newton, Darwin, and Einstein. For a long time it was common for science textbooks to have an early section describing "the scientific method," but recently textbooks seem to have become more cautious about this.

I said that much twentieth-century philosophy of science aimed at describing the *logical* structure of science. What does this mean? The idea is that the philosopher should think of a scientific theory as an abstract structure, something like a set of interrelated sentences. The philosopher aims to give a description of the logical relations between the sentences in the theory and the relations between the theory and observational evidence. Philosophy can also try to describe the logical relations between different scientific theories in related fields.

Philosophers taking this approach tend to be enthusiastic about the tools of mathematical logic. They prize the rigor of their work. This kind of philosophy has often prompted frustration in people working on the actual history and social structure of science. The crusty old philosophers seemed to be deliberately removing their work from any contact with science as it is actually conducted, perhaps in order to hang onto a set of myths about the perfect rationality of the scientific enterprise, or in order to have nothing interfere with the endless games that can be played with imaginary theories expressed in artificial languages. This kind of logic-based philosophy of science will be discussed in the early chapters of this book. I will argue that the logical investigations were often very interesting, but ultimately my sympathy lies with those who insist that philosophy of science should have more contact with actual scientific work.

If looking for a recipe is too simplistic, and looking for a logical theory is too abstract, what might we look for instead? Here is an answer that will be gradually developed as the book goes on: we can try to describe the scientific *strategy* for investigating the world. And we can then hope to describe what sort of *connection* to the world we are likely to achieve by following that strategy. Initially, this may sound vague or impossible, or both. But by the end of the book I hope to show that it makes good sense.

Several times now I have mentioned fields that "neighbor" on philosophy of science—history of science, sociology of science, and parts of psychology, for example. What is the relation between philosophical theories of science and ideas in these neighboring fields? This question was part of the twentieth-century roller-coaster ride that I referred to earlier. Some people in these neighboring fields thought they had reason to believe that the whole idea of a philosophical theory of science is misguided. They expected that philosophy of science would be replaced by fields like sociology. This replacement never occurred. What did happen was that people in these neighboring fields constantly found themselves doing philosophy themselves, whether they realized it or not. They kept running into questions about truth, about justification, and about the connections between theories and reality. The philosophical problems refused to go away.

Philosophers themselves differ a great deal about what kind of input from these neighboring fields is relevant to philosophy. This book is written from a viewpoint that holds that philosophy of science benefits from lots of input from other fields. But the argument that philosophy of science needs that kind of input will not be given until chapter 10.

1.4 Three Answers, or Pieces of an Answer

In this section I will introduce three different answers to our general questions about how science works. In different ways, these three ideas will be recurring themes throughout the book.

The three ideas can be seen as rivals; they can be seen as alternative starting points, or paths into the problem. But they might instead be considered as pieces of a single, more complicated answer. The problem then becomes how to fit them together.

The first of the three ideas is *empiricism*. Empiricism encompasses a diverse family of philosophical views, and debates within the empiricist camp can be intense. But empiricism is often summarized using something like the following slogan:

Empiricism: The only source of real knowledge about the world is experience.

Empiricism, in this sense, is a view about where *all* knowledge comes from, not just scientific knowledge. So how does this help us with the philosophy of science? In general, the empiricist tradition has tended to see the differences between science and everyday thinking as differences of *detail and degree*. The empiricist tradition has generally, though not always, tended to construe science in a broad way, and it has tended to approach questions in the philosophy of science from the standpoint of a general theory of thought and knowledge. The empiricist tradition in philosophy has also been largely *pro*-science; science is seen as the best manifestation of our capacity to investigate and know the world.

So here is a way to use the empiricist principle above to say something about science:

Empiricism and Science: Scientific thinking and investigation have the same basic pattern as everyday thinking and investigation. In each case, the only source of real knowledge about the world is experience. But science is especially successful because it is organized, systematic, and especially responsive to experience.

So "the scientific method," insofar as there is such a thing, will be routinely found in everyday contexts as well. There was no fundamentally *new* approach to investigation discovered during the Scientific Revolution, according to this view. Instead, Europe was freed from darkness and dogmatism by a few brave and brilliant souls who enabled intellectual culture to "come to its senses."

Some readers are probably thinking that these empiricist principles are empty platitudes. *Of course* experience is the source of knowledge about the world—what *else* could be?

For those who suspect that basic empiricist principles are completely trivial, an interesting place to look is the history of medicine. The history of medicine has many examples of episodes where *huge* breakthroughs were made by people willing to make very basic empirical tests—in the face of much skepticism, condescension, and opposition from people who "knew better." Empiricist philosophers have long used these anecdotes to fire up their readers. Carl Hempel, one of the most important empiricist philosophers of the twentieth century, liked to use the sad example of Ignaz Semmelweiss (see Hempel 1966). Semmelweiss worked in a hospital in Vienna in the mid-nineteenth century; he was able to show by simple empirical tests that if doctors washed their hands before delivering babies, the risk of infection in the mothers was hugely reduced. For this radical claim he was opposed and eventually driven out of the hospital.

An even simpler example, which I will describe in some detail to provide a change from the usual case of Semmelweiss, has to do with the discovery of the role of drinking water in the transmission of cholera.

Cholera was a huge problem in cities in the eighteenth and nineteenth centuries, producing death from terrible diarrhea. Cholera is still a problem whenever there are poor people crowded together without good sanitation, as it is transmitted from the diarrhea through drinking water. In the eighteenth and nineteenth centuries, there were various theories of how cholera was caused—this was before the discovery of the role of bacteria and other microorganisms in infectious disease. Some thought the disease was caused by foul gases, called *miasmas,* exuded from the ground and swamps. In London, John Snow hypothesized that cholera was spread by drinking water. He mapped the outbreak of one epidemic in London in 1854 and found that it seemed to be centered on a particular public water pump in Broad Street. With great difficulty he persuaded the local authorities to remove the pump's handle. The outbreak immediately went away.

This was a very important event in the history of medicine. It was central to the rise of the modern emphasis on clean drinking water and sanitation, a movement that has had an immense effect on human health and well-being. This is also the kind of case that shows the attractiveness of even very simple empiricist views.

You might be thinking that we can just end the book here. Empiricism wins; looking to experience is a sure-fire guarantee of getting things right. Those who are tempted to think that no problems remain might consider a cautionary tale that follows up the Snow story. This is the tale of brave Doctor Pettenkofer.

Some decades after Snow, the theory that diseases like cholera are caused by microorganisms—the "germ theory of disease"—was developed in detail by Robert Koch and Louis Pasteur. Koch isolated the bacteria responsible for cholera quite early on. Pettenkofer, however, was unconvinced. To prove Koch wrong, he *drank* a glass of water mixed with the alleged cholera germs. Pettenkofer suffered no ill effects, and he wrote to Koch saying he had disproved Koch's theory.

It is thought that Pettenkofer might have had high stomach acid, which can protect people against cholera infection. Or perhaps the cholera germs had died in that sample. Clearly Pettenkofer was lucky; Koch was right about what causes cholera. But the case reminds us that direct empirical tests are no *guarantee* of success.

Some readers, I said, might be thinking that empiricism is true but too obvious to be interesting. Another line of criticism holds that empiricism is false, because it is committed to an absurdly simple picture of thought, belief, and justification. The empiricist slogan I gave earlier suggests that experiences pour into the mind and somehow turn into knowledge. It turns out to be very difficult to refine basic empiricist ideas in a way that makes them more psychologically realistic. Empiricists do not deny that reasoning, including very elaborate reasoning, is needed to make sense of what we observe. Still, they insist that the role of experience is somehow fundamental in understanding how we learn about the world. Many critics of empiricism hold that this is a mistake; they see it as a hangover from a simplistic and outdated picture of how belief and reasoning work. That debate will be a recurring theme in this book.

I now turn to the second of the three families of views about how science works. This view can be introduced with a quote from Galileo, one of the superheroes of the Scientific Revolution:

> Philosophy is written in this grand book the universe, which stands continually open to our gaze. But the book cannot be understood unless one first learns to comprehend the language and to read the alphabet in which it is composed. *It is written in the language of mathematics,* and its characters are triangles, circles, and other geometric figures without which it is humanly impossible to understand a

single word of it; without these, one wanders about in a dark labyrinth. (Galileo [1623] 1990, 237–38, emphasis added)

Putting the point in plainer language, here is the second of the three ideas.

Mathematics and Science: What makes science different from other kinds of investigation, and especially successful, is its attempt to understand the natural world using mathematical tools.

Is this idea an alternative to the empiricist approach, or something that can be combined with it? Perhaps surprisingly, an emphasis on mathematical methods has often been used to argue against empiricism. Sometimes this has been because people have thought that mathematics shows us that there must be another route to knowledge beside experience; experience is *a* source of knowledge, but not the *only* important source. Alternatively, we might claim that empiricism is trivial: of course knowledge is based on experience, but that tells us nothing about what differentiates science from other human thought. What makes science special is its attempt to quantify phenomena and detect mathematical patterns in the flow of events.

Nonetheless, it is surely sensible to see an emphasis on mathematics as something that can be combined with empiricist ideas. It might seem that Galileo would disagree; Galileo not only exalted mathematics but praised his predecessor Copernicus for making "reason conquer sense [experience]" in his belief that the earth goes around the sun. But this is a false opposition. In suggesting that the earth goes round the sun, Copernicus was not ignoring experience but dealing with apparent conflicts between different aspects of experience, in the light of his background beliefs. And there is no question that Galileo was a very empirically minded person; an emphasis on observations made using the telescope was central to his work, for example. So avoiding the false oppositions, we might argue that mathematics used *as a tool within an empiricist outlook* is what makes science special.

In this book the role of mathematics will be a significant theme but not a central one. This is partly because of the history of the debates surveyed in the book, and partly because mathematical tools are not quite as essential to science as Galileo thought. Although mathematics is clearly of huge importance in the development of physics, one of the greatest achievements in all of science—Darwin's achievement in *On the Origin of Species* ([1859] 1964)—makes no real use of mathematics. Darwin was not confined to the "dark labyrinth" that Galileo predicted as the fate of nonmathematical investigators. In fact, most (though not all) of the huge leaps in biology that occurred in the nineteenth century occurred without much of a role for mathematics. Biology *now* contains many mathematical parts, including modern formulations of Darwin's theory of evolution, but this is a more recent development.

So not all of science—and not all of the greatest science—makes much use of mathematics to understand the world.

The third of the three families of ideas is newer. Maybe the unique features of science are only visible when we look at scientific *communities.*

Social Structure and Science: What makes science different from other kinds of investigation, and especially successful, is its unique social structure.

Some of the most important recent work in philosophy of science has had to do with exploring this idea, but it took the input of historians and sociologists of science to bring philosophical attention to bear on it.

In the hands of historians and sociologists, an emphasis on social structure has often been developed in a way that is strongly critical of the empiricist tradition. Steven Shapin argues that mainstream empiricism often operates within the fantasy that each individual can observationally test hypotheses for himself (Shapin 1994). Empiricism is supposed to urge that people be distrustful of authority and go out to look directly at the world. But of course this is a fantasy. It is a fantasy in the case of everyday knowledge, and it is an even greater fantasy in the case of science. Almost every move that a scientist makes depends on elaborate networks of cooperation and trust. If each individual insisted on testing everything himself, science would never advance beyond the most rudimentary ideas. Cooperation and lineages of transmitted results are essential to science. The case of John Snow and cholera, discussed earlier in this section, is very unusual. Snow looks like a "lone ranger" striding up to the Broad Street water pump (with crowds of empiricists cheering in the background). And even Snow must have been dependent on the testimony of others in his assessment of the state of the cholera epidemic before and after his intervention at the pump.

So trust and cooperation are essential to science. But who can be trusted? Who is a reliable source of data? Shapin argues that when we look closely, a great deal of what went on in the Scientific Revolution had to do with working out new ways of policing, controlling, and coordinating the actions of groups of people in the activity of research. Experience is everywhere. The hard thing is working out which *kinds* of experience are relevant to the testing of hypotheses, and working out who can be trusted as a source of reliable and relevant reports.

So Shapin argues that a good theory of the social organization of science will be a better *theory of science* than empiricist fantasies. But philosophers have begun to develop theories of how science works that emphasize social organization but are also intended to fit in with a form of empiricism (Hull 1988; Kitcher 1993). These accounts of science stress the special balance of cooperation and competition found in scientific communities. People sometimes imagine that seeking individual credit and competition for status and recognition are recent developments in science. But these issues have been important since the time of the Scientific Revolution. The great scientific societies, like the Royal Society of London, came into being quite early—1660 in the case of the Royal Society. A key part of their role was to handle the allocation of credit in an efficient way—making sure the right people were rewarded, without hindering the free spread of ideas. These societies also functioned to create a community of people who could trust each other as reliable co-workers and sources of data. The empiricist can argue that this social organization made scientific *communities* uniquely responsive to experience.

In this section I have sketched three families of ideas about how science works and what makes it distinctive. Each idea has sometimes been seen as *the* starting point for an understanding of science, exclusive of the other two. But it is more likely that they should be seen as pieces of a more complete answer. The first and third ideas—empiricism and social structure—are especially important. These we will return to over and over again. Part of the challenge for philosophy of science in the years to come lies in integrating the insights of the empiricist tradition with the role for social organization in understanding science. That does require significant changes to traditional empiricist ideas.

1.5 Historical Interlude: A Sketch of the Scientific Revolution

Before diving into the philosophical theories, we will take a brief break. Several times already I have mentioned the Scientific Revolution. People, events, and theories from this period carry special weight in discussions of the nature of science. So in this section I will give a historical sketch of the main landmarks, many of which will appear from time to time in later chapters. Before setting out, I should note that there is a good deal of controversy about how to understand this period of history; for example, some historians think that the whole idea of christening this period "The Scientific Revolution" is a mistake, as this phrase makes it sound like there are sharp boundaries between one totally unique period and the rest of history (Shapin 1996). But I will use the phrase in the traditional way.

The Scientific Revolution occurred roughly between 1550 and 1700. These events are positioned at the end of a series of dramatic changes in

Europe, and the Scientific Revolution itself fed into further processes of change. In religion, the Catholic Church had been challenged by Protestantism. The Renaissance of the fifteenth and sixteenth centuries had included a partial opening of intellectual culture. Populations were growing (recovering from the Black Death), and there was increased activity in commerce and trade. Traditional hierarchies, including intellectual hierarchies, were beginning to show strain. As recent writers have stressed, this was a time in which many new, unorthodox ideas were floating around.

The worldview that had been inherited from the Middle Ages was a combination of Christianity with the ideas of the ancient Greek philosopher Aristotle. The combination is often called the Scholastic worldview, after the universities or "Schools" that developed and defended it. The earth was seen as a sphere positioned at the center of the universe, with the moon, sun, planets, and stars revolving around it. A detailed model of the motions of these celestial bodies had been developed by Ptolemy around 150 A.D. (the sun was placed between Venus and Mars).

Aristotle's physical theory distinguished "natural" from "violent" or unnatural motion. The theory of natural motions was part of a more general theory of change in which biological development (from acorn to oak, for example) was a central guiding case, and many events were explained using the idea of *purpose*.

Everything on earth was considered to be made up of mixtures of four basic elements (earth, air, fire, and water), each of which had natural tendencies. Objects containing a lot of earth, for example, naturally fall toward the center of the universe, while fire makes things rise. Unnatural motions, such as the motions of projectiles, have an entirely different kind of explanation. Objects in the heavens are made of a fifth element, which is "incorruptible," or unchanging. The natural motion for objects made of this fifth element is circular.

Some versions of this picture included a mechanism (using the term loosely) for the motions of sun, planets, and stars. For example, each body orbiting the earth might be positioned on a crystalline sphere that revolved around the earth. Ptolemy's own model was harder to interpret in these terms; Ptolemy is sometimes thought to be most interested in giving a tool for astronomical prediction (though interpreters differ on this).

In 1543 the Polish astronomer Nicolaus Copernicus (1473–1543) published a work outlining an alternative picture of the universe. Others had speculated in ancient times that the earth might move around the sun instead of vice versa, but Copernicus was the first to give a detailed theory of this kind. In his theory the earth has two motions, revolving on its axis once a day and orbiting the sun once a year. Copernicus's theory had the same basic placement of the sun, moon, earth, and the known planets that modern astronomy has. But the theory was made more complicated by his insistence, following Aristotle and Ptolemy, that heavenly motions must be circular. Both the Ptolemaic system and Copernicus's system saw most orbits as complex compounds of circles, not single circles. Ptolemy's and Copernicus's systems were about equally complicated, in fact. Writers seem to differ on whether Copernicus's theory was much more accurate as a predictive tool. But there were some famous phenomena that Copernicus's theory explained far better than Ptolemy's. One was the "retrograde motion" of the planets, an apparently erratic motion in which planets seem to stop and backtrack in their motions through the stars.

Copernicus's work aroused interest, but there seemed to be compelling arguments against taking it to be a literally true description of the universe. Some problems were astronomical, and others had to do with obvious facts about motion. Why does an object dropped from a tower fall at the foot of the tower, if the earth has moved a considerable distance while the object is in flight? Copernicus's 1543 book had an extra preface written by a clergyman, Andreas Osiander, who had been entrusted with the publication, urging that the theory be treated just as a calculating tool. This became a historically important statement of a view about the role of scientific theories known as *instrumentalism,* which holds that we should think of theories only as predictive tools rather than as attempts to describe the hidden structure of nature.

The situation was changed dramatically by Galileo Galilei (1564–1642), working in Italy in the early years of the seventeenth century. Galileo vigorously made the case for the literal truth of the Copernican system, as opposed to its mere usefulness. Galileo used telescopes (which he did not invent but did improve) to look at the heavens, and he found a multitude of phenomena that contradicted Aristotle and the Scholastic view of the world. He also used a combination of mathematics and experiment to begin the formulation of a new science of motion that would make sense of the idea of a moving earth and explain familiar facts about dropped and thrown objects. Galileo's work eventually aroused the ire of the pope; he was forced to recant his Copernican beliefs by the Inquisition and spent his last years under house arrest. (Galileo was treated lightly in comparison with Giordano Bruno, whose refusal to disown his unorthodox speculations about the place of the earth in the universe led to his being burned at the stake in Rome, for heresy, in 1600.)

Galileo remained wedded to circular motion as astronomically fundamental. The move away from circular motion was taken by Johannes Kepler (1571–1630), a mystical thinker who combined Copernicanism with an obsession with finding mathematical harmony (including musical tunes) in the structure of the heavens. Kepler's model of the universe, also developed around the start of the seventeenth century, had the earth and other planets moving in *ellipses*, rather than circles, around the sun. This led to massive simplification and better predictive accuracy.

So far I have mentioned only changes in astronomy and related areas of physics, and I have taken the discussion only to the early part of the seventeenth century. Part of what makes this initial period so dramatic is the removal of the earth from the center of the universe, an event laden with symbolism. Another field that changed in the same period is anatomy. In Padua, Andreas Vesalius (publishing, like Copernicus, in 1543) began to free anatomy from dependence on ancient authority (especially Galen's conclusions) and set it on a more empirical path. Influenced by Vesalius's school, William Harvey achieved the most famous breakthrough in this period, establishing in 1628 the circulation of blood and the role of the heart as a pump.

The mid-seventeenth century saw the rise of a general and ambitious new theory about matter: *mechanism*. The mechanical view of the world combined ideas about the composition of things with ideas about causation and explanation. According to mechanism, the world is made up of tiny "corpuscles" of matter, which interact only by local physical contact. Ultimately, good explanations of physical phenomena should only be given in terms of mechanical interactions. The universe was to be understood as operating like a mechanical clock.

Some, like René Descartes (1596–1650), thought that an immaterial soul and a traditional God must be posited as well as physical corpuscles. Though many figures in the Scientific Revolution held religious views that were at least somewhat unorthodox, most were definitely not looking for a showdown with mainstream religion. Most of the "mechanical philosophers" retained a role for a Christian God in their overall pictures of the world. (If the world is a clock, who set it in motion, for example?) However, the idea of dropping souls, God, or both from the picture was sometimes considered.

In England, Robert Boyle (1627–91) and others embedded a version of mechanism into an organized and well-publicized program of research that urged systematic experiment and the avoidance of unempirical speculation. In the mid-seventeenth century we also see the rise of scientific societies in London, Paris, and Florence. These societies were intended to organize the new research and break the institutional monopoly of the (often conservative) universities.

The period ends with the work of Isaac Newton (1642–1727). In 1687 Newton published his *Principia,* which gave a unified mathematical treatment of motion both on earth and in the heavens. Newton showed why Kepler's elliptical orbits were the inevitable outcome of the force of gravity operating between heavenly bodies, and he vastly improved the ideas about motion on earth that Galileo (and others) had pioneered. So impressive was this work that for hundreds of years Newton was seen as having essentially completed those parts of physics. Newton also did immensely influential work in mathematics and optics, and he suggested the way to move forward in fields like chemistry. In some ways Newton's physics was the culmination of the mechanical worldview, but in some ways it was "post-mechanical," since it posited some forces (gravity, most importantly) that were hard to interpret in mechanical terms.

So by the end of the seventeenth century, the Scholastic worldview had been replaced by a combination of Copernicanism and a form of mechanism. As far as method is concerned, a combination of experiment and mathematical analysis had triumphed (though people disagreed about the nature of the triumphant combination). This ends the period usually referred to as the Scientific Revolution. But the changes described above fed into further changes, both intellectual and political. Chemistry began a period of rapid development in the middle to late eighteenth century, a period sometimes called the Chemical Revolution. The work of Lavoisier, especially his description of oxygen and its role in combustion, is often taken to initiate this "revolution," though it was in the nineteenth century, with the work of Dalton, Mendeleyev, and others, that the basic features of modern chemistry, like the periodic table of elements, were established.

Linnaeus had systematized biological classification in the eighteenth century, but it was the nineteenth century that saw dramatic developments in biology. These developments include the theory that organisms are comprised of cells, Darwin's theory of evolution, the germ theory of disease, and the work by Mendel on inheritance that laid the foundation for genetics.

The Scientific Revolution also fed into more general cultural and political changes. In the eighteenth century the philosophers of the French Enlightenment hoped to use science and reason to sweep away ignorance and superstition, along with oppressive religious and political institutions. The intellectual movements leading to the American and French Revolutions in the late eighteenth century were much influenced by currents of thought in science and philosophy. These included empiricism, mechanism, the inspiration of Newton, and a general desire to understand mankind and society in a way modeled on the understanding of the physical world achieved during the Scientific Revolution.

Further Reading

The topics in this chapter will be discussed in detail later, and references will be given then. Two other introductory books are worth mentioning, though. Hempel's *Philosophy of Natural Science* (1966) was for many years the standard introductory textbook in this area. It opens with the story of Semmelweiss and is a clear and reasonable statement of mainstream twentieth-century empiricism. Alan Chalmers's *What Is This Thing Called Science?* (1999) is also very clear; it presents a different view from Hempel's and the one defended here.

For all the topics in this book, there are also reference works that readers may find helpful. Simon Blackburn's *Oxford Dictionary of Philosophy* is a remarkably useful book and is fun to browse through. The *Routledge Encyclopedia of Philosophy* is also of high quality. *The Blackwell Companion to the Philosophy of Science* has many short papers on key topics (though many of these papers are quite advanced). The *Stanford Online Encyclopedia of Philosophy* is still in progress but will be a very useful (and free) resource.

There are many good books on the Scientific Revolution, each with a different emphasis. Cohen, *The Birth of a New Physics* (1985), is a classic and very good on the physics. Henry, *The Scientific Revolution and the Origins of Modern Science* (1997), is both concise and thorough. It has an excellent chapter on mechanism and contains a large annotated bibliography. Schuster 1990 is also a useful quick summary, and Dear's *Revolutionizing the Sciences* (2001) is a concise and up-to-date book with a good reputation. But Toulmin and Goodfield's *Fabric of the Heavens* (1962), an old book recently reprinted, is my favorite. It focuses on the conceptual foundations underlying the development of scientific ideas. (It is the first of three books by Toulmin and Goodfield on the history of science; the second, *The Architecture of Matter* is also relevant here.)

Kuhn's *Copernican Revolution* (1957), is another classic, focused on the early stages, as the title suggests. Shapin's *Scientific Revolution* (1996), is not a good introduction to the Scientific Revolution but is a very interesting book anyway. There are several good books that focus on particular personalities. Koestler, *The Sleepwalkers* (1968), is fascinating on Kepler, and Sobel, *Galileo's Daughter* (1999), is also good on Galileo (and his daughter, a nun leading a tough life). The standard biography of the amazingly strange Isaac Newton, by Robert Westfall, comes in both long (1980) and short (1993) versions.

For a history of medicine, covering the whole world, see Porter, *The Greatest Benefit to Mankind* (1998).

2

Logic Plus Empiricism

2.1 The Empiricist Tradition

The first approach to science that we will examine is a revolutionary form of empiricism that appeared in the early part of the twentieth century, flourished for a time, was transformed and moderated under the pressure of objections, and then slowly became extinct. The earlier version of the view is called "logical positivism," and the later, moderate form is more usually called "logical empiricism." There is variation in terminology here; "logical empiricism" is sometimes used for the whole movement, early and late. Although we will be looking at fossils in this chapter, these remnants of the past are of great importance in understanding where we are now.

Before discussing logical positivism, it will be helpful to go even further back and say something about the empiricist tradition in general. In the first chapter I said that empiricism is often summarized with the claim that the only source of knowledge is experience. This idea goes back a long way, but the most famous stage of empiricist thought was in the seventeenth and eighteenth centuries, with the work of John Locke, George Berkeley, and David Hume. These "classical" forms of empiricism were based upon theories about the mind and how it works. Their view of the mind is often called "sensationalist." Sensations, like patches of color and sounds, appear in the mind and are all the mind has access to. The role of thought is to track and respond to patterns in these sensations. This view of the mind is not implied by the more basic empiricist idea that experience is the source of knowledge, but for many years such a view was common within empiricism.

Both during these classical discussions and more recently, a problem for empiricism has been a tendency to lapse into *skepticism,* the idea that we cannot know anything about the world. This problem has two aspects. One aspect we can call *external world skepticism:* how can we ever know anything about the real world that lies behind the flow of sensations? The second aspect, made vivid by David Hume, is *inductive skepticism:* why do we have reason to think that the patterns in past experience will also hold in the future?

Empiricism has often shown a surprising willingness to throw in the towel on the issue of external world skepticism. (Hume threw in the towel on both kinds of skepticism, but that is unusual.) Many empiricists have been willing to say that they don't *care* about the possibility that there might be real things lying behind the flow of sensations. It's only the sensations that we have any dealings with. Maybe it makes no sense even to try to *think* about objects lying behind sensations. Perhaps our concept of the world is just a concept of a patterned collection of sensations. This view is sometimes called "phenomenalism." During the nineteenth century, phenomenalist views were quite popular within empiricism, and their oddity was treated with nonchalance. John Stuart Mill, an English philosopher and political theorist, once said that matter may be defined as "a Permanent Possibility of Sensation" (1865, 183). Ernst Mach, an Austrian physicist and philosopher, illustrated his phenomenalist view by drawing a picture of the world as it appeared through his left eye (see fig. 2.1; the shape in the lower right part of the image is his elegant mustache). All that exists is a collection of observer-relative sensory phenomena like these.

I hope phenomenalism looks strange to you, despite its eminent proponents. It *is* a strange idea. But empiricists have often found themselves backing into views like this. This is partly because they have often tended to think of the mind as *confined* behind a "veil of ideas" or sensations. The mind has no "access" to anything outside the veil. Many philosophers, including me, agree that this picture of the mind is a mistake. But it is not so easy to set up an empiricist view that entirely avoids the bad influence of this picture.

In discussions of the history of philosophy, it is common to talk of a showdown in the seventeenth and eighteenth centuries between "the rationalists" and "the empiricists." Rationalists like Descartes and Leibniz believed that pure reasoning can be a route to knowledge that does not depend on experience. Mathematics seemed to be a compelling example of this kind of knowledge. Empiricists like Locke and Hume insisted that experience is our only way of finding out what the world is like. In the late eighteenth century, a sophisticated intermediate position was developed by the German philosopher Immanuel Kant. Kant argued that all our thinking involves a subtle *interaction* between experience and preexisting mental structures that we use to *make sense* of experience. Key concepts like space, time, and causation cannot be derived from experience, because a person must *already* have these concepts in order to use experience to learn

Fig. 2.1
"The assertion, then, is correct that the world consists only of our sensations" (Mach 1897, 10).

about the world. Kant also held that mathematics gives us real knowledge of the world but does not require experience for its justification.

Empiricists must indeed avoid overly simple pictures of how experience affects belief. The mind does not passively receive the imprint of facts. The active and creative role of the mind must be recognized. The trick is to avoid this problem while still remaining true to basic empiricist principles.

As I said above, in the history of philosophy the term "rationalism" is often used for a view that opposes empiricism. In the more recent discussions of science that we are concerned with here, however, the term is generally not used in that way. (This can be a source of confusion; see the glossary.) The views called "rationalist" in the twentieth century were often forms of empiricism; the term was often used in a broad way, to indicate a confidence in the power of human reason.

So much for the long history of debate. Despite various problems, empiricism has been a very attractive set of ideas for many philosophers. Empiricism has often also had a particular kind of impact on discussions outside of philosophy. Making a sweeping generalization, it is fair to say that the empiricist tradition has tended to be (1) pro-science, (2) worldly rather than religious, and (3) politically moderate or liberal (though these political labels can be hard to apply across times). David Hume, John Stuart Mill, and Bertrand Russell are examples of this tendency. Of the three elements of my generalization, religion is the one that has the most counterexamples. Berkeley was a bishop, for example, and Bas van Fraassen, one of the most influential living empiricist philosophers, is also religious. But on the whole it is fair to say that empiricist ideas have tended to be the allies of a practical, scientific, down-to-earth outlook on life. The logical positivists definitely fit this pattern.

2.2 The Vienna Circle

Logical positivism was a form of empiricism developed in Europe after World War I. The movement was established by a group of people who were scientifically oriented and who disliked much of what was happening in philosophy. This group has become known as the *Vienna Circle.*

The Vienna Circle was established by Moritz Schlick and Otto Neurath. It was based, as you might expect, in Vienna, Austria. From the early days through to the end, a central intellectual figure was Rudolf Carnap. Carnap seems to have been the kind of person whose presence inspired awe even in other highly successful philosophers.

Logical positivism was an extreme, swashbuckling form of empiricism. The term "positivism" derives from the nineteenth-century scientific philosophy of Auguste Comte. In the 1930s Carnap suggested that they change the name of their movement from "logical positivism" to "logical empiricism." This change should not be taken to suggest that the later stages in the movement were "more empiricist" than the earlier stages. The opposite is true. In my discussion I will use the term "logical positivism" for the intense, earlier version of their ideas, and "logical empiricism" for the later, more moderate version. Although Carnap suggested the name change in the mid-1930s, the time during which logical positivist ideas changed most markedly was after World War II. I will spend some time in this section describing the unusual intellectual and historical context in which logical positivism developed. In particular, it is easier to understand logical positivism if we pay attention to what the logical positivists were *against.*

The logical positivists were inspired by developments in science in the early years of the twentieth century, especially the work of Einstein. They also thought that developments in logic, mathematics, and the philosophy of language had shown a way to put together a new kind of empiricist philosophy that would settle, once and for all, the problems that philosophy had been concerned with. Some problems would be solved, and other problems would be rejected as meaningless. Logical positivist views about language were influenced by the early ideas of Ludwig Wittgenstein ([1922] 1988). Wittgenstein was an enigmatic, charismatic, and eccentric philosopher of logic and language who was not an empiricist at all. Some would say that the positivists adapted Wittgenstein's ideas, others that they misinterpreted him.

Though they did admire some philosophers, the logical positivists were distressed with much of what had been going on in philosophy. In the years after Kant's death in 1804, philosophy had seen the rise of a number of systems of thought that the logical positivists found pretentious, obscure, dogmatic, and politically harmful. One key villain was G. W. F. Hegel, who worked in the early nineteenth century and had a huge influence on nineteenth-century thought. Hegel was famous for his work on the relation between philosophy and history. He thought that human history as a whole was a process in which a "world spirit" gradually reached consciousness of itself. For Hegel, individuals are less important than the state as a whole, especially the role of the state in the grand march of historical progress. These ideas were taken to support strong forms of nationalism. Hegel's was an "idealist" philosophy, since it held that reality is in some sense spiritual or mental. But this is not a view in which each person's reality is made up in some way by that person's ideas. Rather, a single reality *as a whole* is said to have a spiritual or rational character. This view is sometimes called "absolute idealism."

Hegel's influence bloomed and then receded in continental Europe. As it receded in continental Europe, in the later nineteenth century, it bloomed in England and America. Absolute idealism is a good example of what logical positivism was against. Sometimes the positivists would disparagingly dissect especially obscure passages from this literature. Hans Reichenbach (who was not part of the original Vienna Circle but who was a close ally) began his book *The Rise of Scientific Philosophy* (1951) with a quote from Hegel's most famous work on philosophy and history: "Reason is substance, as well as infinite power, its own infinite material underlying all the natural and spiritual life; as also the infinite form, that which sets the material in motion." Reichenbach lamented that a philosophy student, on first reading this passage, would usually think that it was *his* fault—the student's fault—that he did not understand it. The student would then work away until it finally seemed obvious that Reason was substance, as well as infinite power. . . . For Reichenbach, it is entirely *Hegel's* fault that the passage seems to make no sense. It seems to make no sense because whatever factual meaning the claim might be intended to convey has been smothered with misused language.

People sometimes describe the history of this period as if it was a pitched battle between logical positivism and absolute idealism. That is not how things went. In the early twentieth century, there were many kinds of philosophy jostling and wrangling in Europe. There was a "back to Kant" movement going on (as there seems to be now; perhaps this will happen every hundred years). Another philosopher who came to seem an especially important rival to logical positivism was Martin Heidegger.

Earlier I gave a quick summary of Hegel's ideas. It is much harder to do that for Heidegger. Heidegger is sometimes categorized as an existentialist. Perhaps he is the most famously difficult and obscure philosopher who has ever lived. I will borrow the summary reluctantly given by Thomas Sheehan in the entry for Heidegger in the *Routledge Encyclopedia of Philosophy* (1998): "He argues that mortality is our defining moment, that we are thrown into limited worlds of sense shaped by our being-towards-death, and that finite meaning is all the reality we get." Simplifying even more, Heidegger held that we must understand our lives as based, first and foremost, upon practical coping with the world rather than knowledge of it. All our experience is affected by the awareness that we are traveling toward death. And the best thing we can do in this situation is stare it in the face and live an "authentic" life.

This picture of life might seem to make some sense (especially on a bad day). But Heidegger combined his descriptions of how it feels to live in the world with abstract metaphysical speculation; especially notorious are his discussions of the nature of "Nothing." Heidegger also had one point in common with some (though not all) absolute idealists: his opposition to liberal democratic political ideas.

Heidegger was seen as a key rival by the logical positivists. Carnap gave humorous logical dissections of Heidegger's discussions of Nothing in his lectures. Interestingly, recent work has shown that Carnap and Heidegger understood each other better than was once supposed (Friedman 2000).

Logical positivism was a plea for Enlightenment values, in opposition to mysticism, romanticism, and nationalism. The positivists championed reason over the obscure, the logical over the intuitive. The logical positivists were also internationalists, and they liked the idea of a universal and precise language that everyone could use to communicate clearly. Otto Neurath was the member of the group with the strongest political and social interests. He and various others in the group could be described as democratic socialists. They had a keen interest in some movements in art and architecture at the time, such as the Bauhaus movement. They saw this work as assisting the development of a scientific, internationalist, and practical outlook on society (Galison 1990).

The Vienna Circle flourished from the mid-1920s to the mid-1930s. Logical positivist ideas were imported into England by A. J. Ayer in *Language, Truth, and Logic* (1936), a vivid and readable book that conveys the excitement of the time. Under the influence of logical positivism, and the philosophy of G. E. Moore and Bertrand Russell, English philosophy abandoned absolute idealism and returned to its traditional empiricist emphasis, an emphasis it has retained (more or less) ever since.

In continental Europe the story turned out differently. For we have now, remember, reached the 1930s. The development of logical positivism ran straight into the rise of Adolf Hitler.

Many of the Vienna Circle had socialist leanings, some were Jewish, and there were certainly no Nazis. So the logical positivists were persecuted by the Nazis, to varying degrees. The Nazis encouraged and made use of pro-German, anti-liberal philosophers, who also tended to be obscure and mystical. In contrast to the logical positivists, Martin Heidegger joined the Nazi party in 1933 and remained a member throughout the war.

Many logical positivists fled Europe, especially to the United States. Schlick, unfortunately, did not. He was murdered by a deranged former student in 1936. The logical positivists who did make it to the United States were responsible for a great flowering of American philosophy in the years after World War II. These include Rudolf Carnap, Hans Reichenbach, Carl Hempel, and Herbert Feigl. In the United States the strident voice of logical positivists was moderated. Partly this was because of criticisms of their ideas—criticisms from the side of those who shared their general outlook. But the moderation was no doubt partly due to the different intellectual and political climate in the United States. Austria and Germany in the 1930s had been an unusually intense environment for doing philosophy.

2.3 Central Ideas of Logical Positivism

Logical positivist views about science and knowledge were based on a general theory of language; we need to start here, before moving to the views about science. This theory of language featured two main ideas, the *analytic-synthetic distinction* and the *verifiability theory of meaning*.

The analytic-synthetic distinction will probably strike you as bland and obvious, at least at first. Some sentences are true or false simply in virtue of their meaning, regardless of how the world happens to be; these are analytic. A synthetic sentence is true or false in virtue of both the meaning of the sentence *and* how the world actually is. "All bachelors are unmarried" is the standard example of an analytically true sentence. "All bachelors are bald" is an example of a synthetic sentence, in this case a false one. Analytic truths are, in a sense, empty truths, with no factual content. Their truth has a kind of necessity, but only because they are empty.

This distinction had been around, in various forms, since at least the eighteenth century. The terminology "analytic-synthetic" was introduced by Kant. Although the distinction itself looks uncontroversial, it can be made to do real philosophical work. Here is one crucial piece of work the logical positivists saw for it: they claimed that all of mathematics and logic is analytic. This made it possible for them to deal with mathematical knowledge within an empiricist framework. For logical positivism, mathematical propositions do not describe the world; they merely record our conventional decision to use symbols in a particular way. Synthetic claims about the world can be expressed using mathematical language, such as when it is claimed that there are nine planets in the solar system. But proofs and investigations within mathematics itself are analytic. This might seem strange because some proofs in mathematics are very surprising. The logical positivists insisted that once we break down such a proof into small steps, each step will be trivial and unsurprising.

Earlier philosophers in the rationalist tradition had claimed that some things can be known a priori; this means known *independently of experience*. Logical positivism held that the only things that seem to be knowable a priori are analytic and hence empty of factual content.

A remarkable episode in the history of science is important here. For many centuries, the geometry of the ancient Greek mathematician Euclid was regarded as a shining example of real and certain knowledge. Immanuel Kant, inspired by the immensely successful application of Euclidean geometry to nature in Newtonian physics, even claimed that Euclid's geometry (along with the rest of mathematics) is both synthetic and knowable a priori. In the nineteenth century, mathematicians did work out alternative geometrical systems to Euclid's, but they did so as a mathematical exercise, not as an attempt to describe how lines, angles, and shapes work in the actual world. Early in the twentieth century, however, Einstein's revolutionary work in physics showed that a non-Euclidean geometry *is* true of our world. The logical positivists were enormously impressed by this development, and it guided their analysis of mathematical knowledge. The positivists insisted that pure mathematics is analytic, and they broke geometry into two parts. One part is purely mathematical, analytic, and says nothing about the world. It merely describes possible geometrical systems. The other part of geometry is a set of synthetic claims about which geometrical system applies to our world.

I turn now to the other main idea in the logical positivist theory of language, the *verifiability theory of meaning*. This theory applies only to sentences that are not analytic, and it involves a specific kind of "meaning," the kind involved when someone is trying to say something about the world. Here is how the theory was often put: *the meaning of a sentence consists in its method of verification*. That formulation might sound strange (it always has to me). Here is a formulation that sounds more natural: knowing the meaning of a sentence is knowing how to verify it. And here is a key application of the principle: if a sentence has no possible method of verification, it has no meaning.

By "verification" here, the positivists meant verification *by means of observation*. Observation in all these discussions is construed broadly, to include all kinds of sensory experience. And "verifiability" is not the best word for what they meant. A better word would be "testability." This is because testing is an attempt to work out whether something is true *or* false, and that is what the positivists had in mind. The term "verifiable" generally only applies when you are able to show that something is true. It would have been better to call the theory "the testability theory of meaning." Sometimes the logical positivists did use that phrase, but the more standard name is "verifiability theory," or just "verificationism."

Verificationism is a strong empiricist principle; experience is the only source of meaning, as well as the only source of knowledge. Note that verifiability here refers to verifiability in *principle*, not in practice. There was some dispute about which hard-to-verify claims are really verifiable in principle. It is also important that *conclusive* verification or testing was not required. There just had to be the possibility of finding observational evidence that would count for or against the proposition in question.

In the early days of logical positivism, the idea was that in principle one could *translate* all sentences with factual meaning into sentences that referred only to sensations and the patterns connecting them. This program of translation was fairly quickly abandoned as too extreme. But the verifiability theory was retained after the program of translation had been dropped.

The verifiability principle was used by the logical positivists as a philosophical weapon. Scientific discussion, and most everyday discussion, consists of verifiable and hence meaningful claims. Some other parts of language are clearly not intended to have factual meaning, so they fail the verifiability test but do so in a harmless way. Included are poetic language, expressions of emotion, and so on. But there are also parts of language that are *supposed* to have factual meaning—are supposed to say something about the world—but which *fail* to do so. For the logical positivists, this includes most traditional philosophy, much of ethics, and theology as well!

This analysis of language provided the framework for the logical positivist philosophy of science. Science itself was seen as just a more complex and sophisticated version of the sort of thinking, reasoning, and problem-solving that we find in everyday life—and completely *un*like the meaningless blather of traditional philosophy.

So let us now look at the logical positivists' picture of science and of the role of philosophy in a scientific worldview. Next we should turn to another distinction they made, between "observational" language and "theoretical" language. There was uncertainty about how exactly to set this distinction up. Usually it was seen as a distinction applied to individual terms. "Red" is in the observational part of language, and "electron" is in the theoretical part. There was also a related distinction at the level of sentences. "The rod is glowing red" is observational, while "Helium atoms each contain two electrons" is theoretical. A more important question was where to draw the line. Schlick thought that only terms referring to sensations were observational; everything else was theoretical. Here Schlick stayed close to traditional empiricism. Neurath thought this was a mistake and argued that terms referring to many ordinary physical objects are in the observational part of language. For Neurath, scientific testing must not be understood in a way that makes it private to the individual. Only observation statements about physical objects can be the basis of public or "intersubjective" testing.

The issue became a constant topic of discussion. In time, Carnap came to think that there are lots of acceptable ways of marking out a distinction between the observational and theoretical parts of language; one could use whichever is convenient for the purposes at hand. This was the start of a more general move that Carnap made toward a view based on the "tolerance" of alternative linguistic frameworks.

We now need to look at logical positivist views about logic. For logical positivism, *logic is the main tool for philosophy,* including philosophical discussion of science. In fact, just about the only useful thing that philosophers can do is give logical analyses of how language, mathematics, and science work.

Here we should distinguish two kinds of logic (this discussion will be continued in chapter 3). Logic in general is the attempt to give an abstract theory of what makes some arguments compelling and reliable. Deductive logic is the most familiar kind of logic, and it describes patterns of argument that transmit truth with certainty. These are arguments with the feature that if the premises of the argument are true, the conclusion must be true. Impressive developments in deductive logic had been under way since the late nineteenth century and were still going on at the time of the Vienna Circle.

The logical positivists also believed in a second kind of logic, a kind that was (and is) much more controversial. This is *inductive* logic. Inductive logic was supposed to be a theory of arguments that provide support for their conclusions but do not give the kind of guarantee found in deductive logic.

From the logical positivist point of view, developing an inductive logic was of great importance. Hardly any of the arguments and evidence that we confront in everyday life and science carry the kind of guarantees found in deductive logic. Even the best kind of evidence we can find for a scientific theory is not completely decisive. There is always the possibility of error, but that does not stop some claims in science from being supported by evidence. The logical positivists accepted and embraced the fact that error is always possible. Although some critics have misinterpreted them on this point, the logical positivists did *not* think that science ever reaches absolute certainty.

The logical positivists saw the task of logically analyzing science as sharply distinct from any attempt to understand science in terms of its history or psychology. Those are empirical disciplines, and they involve a different set of questions from those of philosophy.

A terminology standardly used to express the separations between different approaches here was introduced by Hans Reichenbach. Reichenbach distinguished between the "context of discovery" and the "context of justification." That terminology is not helpful, because it suggests that the distinction has to do with "before and after." It might seem that the point being made is that discovery comes first and justification comes afterward. That is not the point being made (though the logical positivists were not completely clear on this). The key distinction is between the study of the logical structure of science and the study of historical and psychological aspects of science.

So logical positivism tended to dismiss the relevance of fields like history and psychology to the philosophy of science. In time this came to be regarded as a big mistake.

Let us put all these ideas together and look at the picture of science that results. Logical positivism was a revolutionary, uncompromising version of empiricism, based largely on a theory of language. The aim of science—and the aim of everyday thought and problem-solving as well—is to track and anticipate patterns in experience. As Schlick once put it, "what every scientist seeks, and seeks alone, are . . . the rules which govern the connection of experiences, and by which alone they can be predicted" (1932–33, 44). We can make rational predictions about future experiences by attending to patterns in past experience, but we never get a guarantee. We could always be wrong. There is no alternative route to knowledge besides experience;

when traditional philosophy has tried to find such a route, it has lapsed into meaninglessness.

The interpretation of logical positivism I have just given is a standard one. There is controversy about how to interpret the aims and doctrines of the movement, however. Some recent writers have argued that there is less of a link between logical positivism and traditional empiricism than the standard interpretation claims (Friedman 1999). But in the sense of empiricism used in this book, there is definitely a strong link. We see that in the Schlick quote given in the previous paragraph.

During the early twentieth century, there were various other strong versions of empiricism being developed as well. One was *operationalism,* which was developed by a physicist, Percy Bridgman (1927). Operationalism held that scientists should use language in such a way that all theoretical terms are tied closely to direct observational tests. This is akin to logical positivism, but it was expressed more as a proposed *tightening up* of scientific language (motivated especially by the lessons of Einstein's theory of relativity) than as an analysis of how all science already works.

In the latter part of the twentieth century, an image of the logical positivists developed in which they were seen as stodgy, conservative, unimaginative science-worshipers. Their strongly pro-science stance has even been seen as antidemocratic, or aligned with repressive political ideas. This is very unfair, given their actual political interests and activities. Later we will see how ideas about the relation between science and politics changed through the twentieth century in a way that made this interpretation possible. The accusation of stodginess is another matter; the logical positivists' writings were often extremely dry and technical. Still, even the driest of their ideas were part of a remarkable program that aimed at a massive, transdisciplinary, intellectual housecleaning. And their version of empiricism was organized around an ideal of intellectual flexibility as a mark of science and rationality. We see this in a famous metaphor used by Neurath (who exemplifies these themes especially well). Neurath said that in our attempts to learn about the world and improve our ideas, we are "like sailors who have to rebuild their ship on the open sea." The sailors replace pieces of their ship plank by plank, in a way that eventually results in major changes but which is constrained by the need to keep the ship afloat during the process.

2.4 Problems and Changes

Logical positivist ideas were always in a state of flux, and they were subject to many challenges. One set of problems was internal to the program. For example, there was considerable difficulty in getting a good formulation of the verifiability principle. It turned out to be hard to formulate the principle in a way that would exclude all the obscure traditional philosophy but include all of science. Some of these problems were almost comically simple. For example, if "Metals expand when heated" is testable, then "Metals expand when heated and the Absolute Spirit is perfect" is also testable. If we could empirically show the first part of the claim to be false, then the whole claim would be shown false, because of the logic of statements containing "and." (If *A* is false then *A*&*B* must be false too.) Patching this hole led to new problems elsewhere; the whole project was quite frustrating (Hempel 1965, chap. 4). The attempt to develop an inductive logic also ran into serious trouble. That topic will be covered in the next chapter.

Other criticisms were directed not at the details but at the most basic ideas of the movement. The criticism that I will focus on here is one of these, and its most famous presentation is in a paper sometimes regarded as the most important in all of twentieth-century philosophy: W. V. Quine's "Two Dogmas of Empiricism" (1953).

Quine argued for a *holistic* theory of testing, and he used this to motivate a holistic theory of meaning as well. In describing the view, first I should say something about holism in general. Many areas of philosophy contain views that are described using the term "holism." A holist argues that you cannot understand a particular thing without looking at its place in a larger whole. In the case we are concerned with here, holism about testing says that we cannot test a single hypothesis or sentence in isolation. Instead, we can only test complex networks of claims and assumptions. This is because only a complex network of claims and assumptions makes definite predictions about what we should observe.

Let us look more closely at the idea that individual claims about the world cannot be tested in isolation. The idea is that in order to test one claim, you need to make assumptions about many other things. Often these will be assumptions about measuring instruments, the circumstances of observation, the reliability of records and of other observers, and so on. So whenever you think of yourself as testing a single idea, what you are really testing is a long, complicated *conjunction* of statements; it is the whole conjunction that gives you a definite prediction. If a test has an unexpected result, then something in that conjunction is false, but the failure of the test itself does not tell you *where* the error is.

For example, suppose you want to test the hypothesis that high air pressure is associated with fair, stable weather. You make a series of observations, and what you seem to find is that high pressure is instead associated with unstable weather. It is natural to suspect that your original hypothesis was wrong, but there are other possibilities as well. It might be that your barometer does not give reliable measurements of air pressure. There might also be something wrong with the observations made (by you or others) of the weather conditions themselves. The unexpected observations are telling you that *something* is wrong, but the problem might lie with one of your background assumptions, not with the hypothesis you were trying to test.

Some parts of this argument are convincing. It is true that only a network of claims and assumptions, not a single hypothesis alone, tells us what we should expect to observe. The failure of a prediction will always have a range of possible explanations. In that sense, testing is indeed holistic. But this leaves open the possibility that we might often have good reasons to lay the blame for a failed prediction at one place rather than another. In practice, science seems to have some effective ways of working out where to lay the blame. Giving a philosophical theory of these decisions is a difficult task, but the mere fact that failed predictions always have a range of possible explanations does not settle the holism debate.

Holist arguments had a huge effect on the philosophy of science in the middle of the twentieth century. Quine, who sprinkled his writings with deft analogies and dry humor, argued that mainstream empiricism had been committed to a badly simplistic view of testing. We must accept, as Quine said in a famous metaphor, that our theories "face the tribunal of sense-experience . . . as a corporate body" (1953, 41). Logical positivism must be replaced with a holistic version of empiricism.

But there is a puzzle here. The logical positivists *already accepted* that testing is holistic in the sense described above. Here is Herbert Feigl, writing in 1943: "No scientific assumption is testable in complete isolation. Only whole complexes of inter-related hypotheses can be put to the test" (1943, 16). Carnap had been saying the same thing (1937, 318). We can even find statements like this in Ayer's *Language, Truth, and Logic* (1936).

Quine did recognize Pierre Duhem, a much earlier French physicist and philosopher, as someone who had argued for holism about testing. (Holism about testing is often called "the Duhem-Quine thesis.") But how could it be argued that logical positivists had dogmatically missed this important fact, when they repeatedly expressed it in print? Regardless of this, many philosophers agreed with Quine that logical positivism had made a bad mistake about testing in science.

Though the history of the issue is strange, it might be fair to say this: although the logical positivists officially accepted a holistic view about testing, they did not appreciate the significance of the point. The verifiability principle *seems* to suggest that you can test sentences one at a time. It seems to attach a set of observable outcomes of tests to each sentence in isolation.

Strictly, the positivists generally held that these observations are only associated with a specific hypothesis *against a background of other assumptions*. But then it seems questionable to associate the test results solely with the hypothesis itself. Quine, in contrast, made the consequences of holism about testing very clear. He also drew conclusions about language and meaning; given the link between testing and meaning asserted by logical positivism, holism about testing leads to holism about meaning. And holism about meaning causes problems for many logical positivist ideas.

The version of holism that Quine defended in "Two Dogmas" was an extreme one. It included an attack on the one idea in the previous section that you might have thought was completely safe: the analytic-synthetic distinction. Quine argued that this distinction *does not exist;* this is another unjustified "dogma" of empiricism.

Here again, some of Quine's arguments were directed at a version of the analytic-synthetic distinction that the logical positivists no longer held. Quine said that the idea of analyticity was intended to treat some claims as *immune to revision,* and he argued that in fact no statement is immune to revision. But Carnap had already decided that analytic statements can be revised, though they are revised in a special way. A person or community can decide to drop one whole linguistic and logical framework and adopt another. Against the background provided by a given linguistic and logical framework, some statements will be analytic and hence not susceptible to empirical test. But we can always change frameworks. By the time that Quine was writing, Carnap's philosophy was based on a distinction between changes made *within* a linguistic and logical framework, and changes *between* these frameworks.

In another (more convincing) part of his paper, Quine argued that there is no way to make *scientific sense* of a sharp analytic-synthetic distinction. He connected this point to his holism about testing. For Quine, all our ideas and hypotheses form a single "web of belief," which has contact with experience only as whole. An unexpected observation can prompt us to make a great variety of possible changes to the web. Even sentences that might look analytic can be revised in response to experience in some circumstances. Quine noted that strange results in quantum physics had suggested to some that revisions in logic might be needed.

In this discussion of problems for logical positivism, I have included some discussions that started early and some that took place after World War II, when the movement had begun its U.S.-based transformation. Let us now look at some central ideas of logical empiricism, the later, less aggressive stage of the movement.

2.5 Logical Empiricism

Let's see how things looked in the years after World War II. Schlick is dead, and other remnants of the Vienna Circle are safely housed in American universities—Carnap at Chicago, Hempel at Pittsburgh and then Princeton, Reichenbach at UCLA (via Turkey), Feigl at Minnesota. Many of the same people are involved, but the work is different. The revolutionary attempt to destroy traditional philosophy has been replaced by a program of careful logical analysis of language and science. Discussion of the contributions that could be made by the scientific worldview to a democratic socialist future have been dropped or greatly muted. (Despite this, the FBI collected a file on Carnap as a possible Communist sympathizer.)

As before, ideas about language guided logical empiricist ideas about science. The analytic-synthetic distinction had not been rejected, but it was regarded as questionable. The logical empiricists felt the pressure of Quine's arguments. The verifiability theory, which had been so scythe-like in its early forms, was replaced by a *holistic empiricist theory of meaning*. Theories were seen as abstract structures that connect many hypotheses together. These structures are connected, as wholes, to the observable realm, but each *bit* of a theory—each claim or hypothesis or concept—does not have some specific set of observations associated with it. A theoretical term (like "electron" or "gene") derives its meaning from its place in the whole structure and from the structure's connection to the realm of observation.

Late in the logical empiricist era, in 1970, Herbert Feigl gave a pictorial representation of what he called "the orthodox view" of theories (see fig. 2.2). A network of theoretical hypotheses ("postulates") is connected by stages to what Feigl calls the "soil" of experience. This anchoring is the source of the network's meaning. Feigl used this picture to describe a single scientific theory. For the more extreme holism of Quine, a person's *total* set of beliefs form a *single* network.

The logical positivist distinction between observational and theoretical parts of language was kept roughly intact. But the idea that observational language describes private sensations had been dropped. The observational base of science was seen as made up of descriptions of observable physical objects (though Carnap thought it might occasionally be useful to work with a language referring to sensations).

Logical positivist views about the role of logic in philosophy and about the sharp separation between the logic of science and the historical-psychological side of science were basically unchanged. A good example of the kind of work done by logical empiricists is provided by their work on

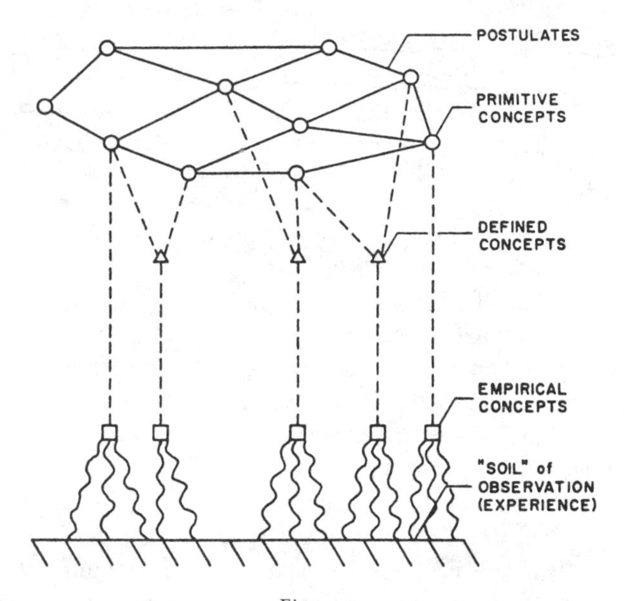

Fig. 2.2
Feigl's picture of the logical empiricist view of theories
(From Feigl 1970; reproduced courtesy
of University of Minnesota Press)

explanation in science (see especially Hempel and Oppenheim 1948; Hempel 1965). For Hempel, to explain something is to show how to *infer* it using a logical argument, where the premises of the argument include at least one statement of a natural law (see chapter 13 below). This illustrates the idea, common to logical positivism and logical empiricism, that logic is the main tool of philosophy of science.

We saw that logical positivism held that the sole aim of science is to track patterns in experience. For logical positivism, when a scientist seems to be trying to describe unobservable structures in the world that give rise to what we see, the scientist must instead be seen as describing the *observable* world in a special, abstract way. Scientific language is only meaningful insofar as it picks out patterns in the flow of experience. Now, does logical *empiricism* make the same claim? Does logical empiricism claim that scientific language ultimately only describes patterns in observables?

The answer is that logical empiricists agonized over this. In their hearts their answer was *yes,* but this answer seemed to get harder and harder to defend. Carl Hempel wrote a paper in 1958 called "The Theoretician's Dilemma," which was the height of logical empiricist agony over the issue. As a fairly traditional empiricist, Hempel was attracted to the idea that the *only possible role* for those parts of language that seem to refer to unobservable entities is to help us pick out patterns in the observable realm. And if the parts of theories that appear to posit unobservable things are really any good, this "goodness" has to show up in advantages the theory has in its handling of observables. So there is no justification for seeing these parts of scientific language as describing real objects lying beyond experience. But Hempel and the logical empiricists found themselves forced to concede that this view does not make much sense of actual scientific work. When scientists use terms like "electron" or "gene," they act as if they are doing more than tracking complex patterns in the observable realm. But the idea that the logical empiricists were being pushed toward—the idea that scientific theories are aimed at describing unobservable real structures—was hard to put on the table and defend. Empiricist philosophy of language seemed implacably opposed to it.

Empiricists were familiar with bad versions of the idea that behind the ordinary world of observables there is a special and superior realm, pure and perfect. This "layered" view of reality seemed to empiricists a source of endless trouble, right from the time of the ancient Greek philosopher Plato, who distinguished the illusory, unstable world of "appearances" from the more perfect and real world of "forms." Empiricists have rightly been determined to avoid this kind of picture. But much of science does appear to be a process in which people hypothesize hidden structures that give rise to observable phenomena. These hidden structures are not "pure and perfect" or "more real" than the observable parts of the world, but they do lie behind or beneath observable phenomena. Of course, unobservable structures posited by a theory at one time might well turn out to be observable at a later time. In science, there is no telling what kinds of new access to the hidden parts of the world we might eventually achieve. But still, much of science does seem to proceed by positing entities that are, at the time of the research in question, truly hidden. For the traditional empiricist philosopher, understanding scientific theorizing in a way that posits a layer of observable phenomena and a layer of hidden structure responsible for the phenomena takes us *far too close* to bad old philosophical views like Plato's. We are too close for comfort, so we must give a different kind of description of how science works.

The result is the traditional empiricist insistence that, ultimately, the only thing scientific language can do is describe patterns in the observable realm. In the first published paper that introduced logical positivism, Carnap, Hahn, and Neurath said: "In science there are no 'depths'; there is surface everywhere" ([1929] 1973, 306). This is a vivid expression of the empiricist aversion to a view in which the aim of theorizing is to describe hidden levels of structure. Science uses unusual theoretical concepts (which *look* initially like attempts to refer to hidden things) as a way of discovering and describing subtle patterns in the observable realm. So the logical positivists and the logical empiricists talked constantly about *prediction* as the goal of science. Prediction was a substitute for the more obvious-looking—but ultimately forbidden—goal of describing the real hidden structure of the world.

Twentieth-century empiricism made an important mistake here. We can make sense of science only by treating much of it as an attempt to describe hidden structures that give rise to observable phenomena. This is a version of *scientific realism*, an idea that will be discussed later in this book. In science there *are* depths. There is not a simple and fixed distinction between two "layers" in nature—the empiricists were right to distrust this idea. Instead there are *many* layers, or rather a *continuum* between structures that are more accessible to us and structures that are less accessible. Genes are hidden from us in some ways, but not as hidden as electrons, which in turn are not as hidden as quarks. Although there are "depths" in science, what is deep at one time can come to the surface at later times, and there may be lots of ways of interacting with what is presently deep.

2.6 On the Fall of Logical Empiricism

Logical empiricist ideas dominated much American philosophy, and they were very influential elsewhere in the English-speaking world and in some parts of Europe, in the middle of the twentieth century. But by the mid-1960s the view was definitely under threat; and by the middle or late 1970s, logical empiricism was near to extinction. The fall of logical empiricism was due to several factors, all of which I have either introduced in this chapter or will discuss in later chapters. One is the breakdown of the view of language that formed the basis of many logical positivist and logical empiricist ideas. Another is pressure from holist arguments. A third is the frustrating history of attempts to develop an inductive logic (chapter 3). A fourth is the development of a new role for fields like history and psychology in the philosophy of science (chapters 5–7). And eventually there was pressure from scientific realism. But this was only possible after logical empiricism had begun to decline.

Further Reading

For much more on the empiricist tradition in general, see Garrett and Barbanell, *Encyclopedia of Empiricism* (1997).

Schlick's "Positivism and Realism" (1932–33) and Feigl's "Logical Empiricism" (1943) are good statements of logical positivism by original members of the Vienna Circle. (Feigl uses the term "logical empiricism," but his paper describes a fairly strong, undiluted version of the view.) Ayer's *Language, Truth, and Logic* (1936) is readable, vivid, and exciting. Some see it as a distortion of logical positivist ideas.

The *Routledge Encyclopedia of Philosophy* (1998) has an interesting collection of articles, especially in the light of new debates about the history of logical positivism. The article on logical positivism is by Friedman and reflects his somewhat unorthodox reading (de-emphasizing the empiricist tradition). Stadler's entry on the Vienna Circle gives a more traditional view. See also Creath's entry on Carnap. On all these issues, see also the essays in Giere and Richardson 1997.

Peter Galison's "Aufbau/Bauhaus" (1990) is a wonderful account of the artistic, social, and political interests of the logical positivists and the links between these interests and their philosophical ideas. Passmore 1966 is a good and accessible survey of philosophical movements and trends in the late nineteenth and early twentieth centuries, including absolute idealism.

Hempel, *Aspects of Scientific Explanation* (1965), is the definitive statement of logical empiricism. His *Philosophy of Natural Science* (1966) is the easy version. Carnap's later lectures have been published as *Introduction to the Philosophy of Science* (1995).

An attempt to revive some logical positivist ideas has recently begun; see, for example, Elliott Sober's forthcoming book *Learning from Logical Positivism.*

3

Induction and Confirmation

3.1 The Mother of All Problems

In this chapter we begin looking at a very important and difficult problem, the problem of understanding how observations can *confirm* a scientific theory. What connection between an observation and a theory makes that observation *evidence for* the theory? In some ways, this has been *the* fundamental problem in the last hundred years of philosophy of science. This problem was central to the projects of logical positivism and logical empiricism, and it was a source of constant frustration for them. And although some might be tempted to think so, this problem does not disappear once we give up on logical empiricism. The problem, in some form or other, arises for nearly everyone.

The aim of the logical empiricists was to develop a *logical* theory of evidence and confirmation, a theory treating confirmation as an abstract relation between sentences. It has become fairly clear that their approach to the problem is doomed. The way to analyze testing and evidence in science is to develop a different kind of theory. But it will take a lot of discussion, in this and later chapters, before the differences between approaches that will and will not work in this area can emerge. The present chapter will mostly look at how the problem of confirmation was tackled in the middle of the twentieth century. And that is a tale of woe.

Before looking at twentieth-century work on these issues, we must again look further into the past. The confirmation of theories is closely connected to another classic issue in philosophy: *the problem of induction*. What reason do we have for expecting patterns observed in our past experience to hold also in the future? What justification do we have for using past observations as a basis for generalization about things we have not yet observed?

The most famous discussions of induction were written by the eighteenth-century Scottish empiricist David Hume ([1739] 1978). Hume asked, What reason do we have for thinking that the future will resemble the past? There is no *contradiction* in supposing that the future could be totally unlike the past. It is *possible* that the world could change radically at any point, rendering previous experience useless. How do we know this will not happen? We might say to Hume that when we have relied on past experience before, this has turned out well for us. But Hume replies that this is begging the question—presupposing what has to be shown. Induction has worked in the past, sure, but that's the *past*! We have successfully used "past pasts" to tell us about "past futures." But our problem is whether *anything* about the past gives us good information about what will happen *tomorrow.*

Hume concluded that we have no reason to expect the past to resemble the future. Hume was an "inductive skeptic." He accepted that we all use induction to make our way around the world. And he was not suggesting that we stop doing so (even if we could). Induction is psychologically natural to us. Despite this, Hume thought it had no rational basis. Hume's inductive skepticism has haunted empiricism ever since. The problem of confirmation is not the same as the classical problem of induction, but it is closely related.

3.2 Induction, Deduction, Confirmation, and Explanatory Inference

The logical empiricists tried to show how observational evidence could provide support for a scientific theory. The idea of "support" is important here; there was no attempt to show that scientific theories could be *proved.* Error is always possible, but evidence can support one theory over another.

The cases that were to be covered by this analysis included the simplest and most traditional cases of induction: if we see a multitude of cases of white swans, and no other colors, why does that give us reason to believe that all swans are white? But obviously not all cases of evidence in science are like this. The observational support for Copernicus's theory that the earth goes around the sun, or for Darwin's theory of evolution, seems to work very differently. Darwin did *not* observe a set of individual cases of evolution and then generalize.

The logical empiricists wanted a theory of evidence, or "theory of confirmation," that would cover all these cases. They were not trying to develop a *recipe* for confirming theories. Rather, the aim was to give an account of the relationships between the statements that make up a scientific theory and statements describing observations, which make the observations support the theory. You might wonder, at this point, what use there could be for a theory with so distant a relationship to actual scientific behavior. Who cares whether a logical analysis of this kind exists or not? In defense of logical empiricism, we might say this: although scientific behavior is not being directly described by the theory of confirmation, nonetheless scientific procedures might be *based* on assumptions described in the theory of confirmation. Perhaps scientists do many things that cannot be justified if confirmation does not exist.

Let us look more closely at what the logical empiricists tried to do. First, I should say more about the distinction between deductive and inductive logic (a distinction introduced in chapter 2). Deductive logic is the well-understood and less controversial kind of logic. It is a theory of patterns of argument that transmit truth with certainty. These arguments have the feature that *if* the premises of the argument are true, the conclusion is guaranteed to be true. An argument of this kind is *deductively valid*. The most famous example of a logical argument is a deductively valid argument:

PREMISES	All men are mortal.
	Socrates is a man.
CONCLUSION	Socrates is mortal.

A deductively valid argument might have false premises. In that case the conclusion might be false as well (although it also might not be). What you get out of a deductive argument depends on what you put in.

The logical empiricists loved deductive logic, but they realized that it could not serve as a complete analysis of evidence and argument in science. Scientific theories do have to be logically *consistent*, but this is not the whole story. Many inferences in science are not deductively valid and give no guarantee. But they still can be *good* inferences; they can still provide *support* for their conclusions.

For the logical empiricists, there is a reason why so much inference in science is not deductive. As empiricists, they believed that all our evidence derives from observation. Observations are always of *particular* objects and occurrences. But the logical empiricists thought that the great aim of science is to discover and establish *generalizations*. Sometimes the aim was seen as describing "laws of nature," but this concept was also regarded with some suspicion. The key idea was that science aims at formulating and testing generalizations, and these generalizations were seen as having an infinite range of application. No finite number of observations can conclusively establish a generalization of this kind, so these inferences from observations in support of generalizations are always nondeductive. (In contrast, all it takes is *one* case of the right kind to prove a generalization to be *false;* this fact will loom large in the next chapter.)

In many discussions of these topics, the logical empiricists (and some later writers) used a simple terminology in which all arguments are either deductive or inductive. Inductive logic was thought of as a theory of *all* good arguments that are not deductive. Carnap, especially, used "induction" in a very broad way. But this terminology can be misleading, and I will set things up differently.

I will use the term "induction" only for inferences from particular observations in support of generalizations. To use the most traditional example, the observation of a large number of white swans (and no swans of any other color) might be used to support the hypothesis that all swans are white. We could express the premises with a list of particular cases—"Swan 1 observed at time t_1 was white; swan 2 observed at time t_2 was white. . . ." Or we might simply say: "All the many swans observed so far have been white." The conclusion will be the claim that all swans are white—a conclusion that could well be false but which is supported, to some extent, by the evidence. Sometimes "enumerative induction" or "simple induction" is used for inductive arguments of this most traditional and familiar kind. Not all inferences from observations to generalizations have this very simple form, though. (And a note to mathematicians: *mathematical induction* is really a kind of *de*duction, even though it has the superficial form of *in*duction.)

A form of inference closely related to induction is *projection.* In a projection, we infer from a number of observed cases to arrive at a prediction about the *next* case, not to a generalization about all cases. So we see a number of white swans and infer that the next swan will be white. Obviously there is a close relationship between induction and projection, but (surprisingly, perhaps) there are a variety of ways of understanding this relationship.

Clearly there are other kinds of nondeductive inference in science and everyday life. For example, during the 1980s Luis and Walter Alvarez began claiming that a huge meteor had hit the earth about 65 million years ago, causing a massive explosion and dramatic weather changes that coincided with the extinction of the dinosaurs (Alvarez et al. 1980). The Alvarez team claimed that the meteor caused the extinctions, but let's leave that aside here. Consider just the hypothesis that a huge meteor hit the earth 65 million years ago. A key piece of evidence for this hypothesis is the presence of unusually high levels of some rare chemical elements, such as iridium, in layers in the earth's crust that are about 65 million years old. These chemical elements tend to be found in meteors in much higher concentrations than they are near the surface of the earth. This observation is taken to be strong evidence supporting the Alvarez theory that a meteor hit the earth around that time.

If we set this case up as an argument, with premises and a conclusion, it clearly is not an induction or a projection. We are not inferring to a generalization, but to a hypothesis about a structure or process that would explain the data. A variety of terms are used in philosophy for inferences of this kind. C. S. Peirce called these "abductive" inferences as opposed to inductive ones. Others have called them "explanatory inductions," "theoretical inductions," or "theoretical inferences." More recently, many philosophers have used the term "inference to the best explanation" (Harman 1965; Lipton 1991). I will use a slightly different term—"explanatory inference."

So I will recognize two main kinds of nondeductive inference, *induction* and *explanatory inference* (plus *projection,* which is closely linked to induction). The problem of analyzing confirmation, or the problem of analyzing evidence, includes all of these.

How are these kinds of inference related to each other? For logical positivism and logical empiricism, induction is the most fundamental kind of nondeductive inference. Reichenbach claimed that all nondeductive inference in science can be reconstructed in a way that depends only on a form of inference that is close to traditional induction. What looks like an explanatory inference can be somehow broken down and reconstructed as a complicated network of inductions and deductions. Carnap did not make this strong claim, but he did seem to view induction as a *model* for all other kinds of nondeductive inference. Understanding induction was in some sense the key to the whole problem. And the majority of the logical empiricist literature on these topics was focused on induction rather than explanatory inference.

So one way to view the situation is to see induction as fundamental. But it is also possible to do the opposite, to claim that explanatory inference is fundamental. Gilbert Harman argued in 1965 that inductions are justified only when they are explanatory inferences in disguise, and others have followed up this idea in various ways.

Explanatory inference seems much more common than induction within actual science. In fact, you might be wondering whether science contains *any* inductions of the simple, traditional kind. That suspicion is reasonable, but it might go too far. Science does contain inferences that look like traditional inductions, at least on the face of them. Here is one example. During the work that led to the discovery of the structure of DNA by James Watson and Francis Crick, a key piece of evidence was provided by "Chargaff's rules." These "rules," described by Erwin Chargaff in 1947, have to do with the relation between the amounts of the four "bases," C, A, T, and G, that help make up DNA. Chargaff found that in the DNA samples he analyzed, the amounts of C and G were always roughly the same, and the amounts of T and A were always roughly the same. This fact about DNA became important in the discussions of how DNA molecules are put together. I called it a "fact" just above, but of course Chargaff in 1947 had not observed all the molecules of DNA that exist, and neither have we. In 1947 Chargaff's claim rested on an induction from a small number of cases (in just eight different kinds of organisms). Today we can give an argument for why Chargaff's rules hold that is not just a simple induction; the structure of DNA explains why Chargaff's rules must hold. But it might appear that, back when the rules were originally discovered, the only reason to take the rules to describe all DNA was inductive.

So it might be a good idea to refuse to treat one of these kinds of inference as "more fundamental" than the other. Maybe there is more than one kind of good nondeductive inference (and perhaps there are others besides the ones I have mentioned). Philosophers often find it attractive to think that there is ultimately just one kind of nondeductive inference, because that seems to be a simpler situation. But the argument from simplicity is unconvincing.

Let us return to our discussion of how the problem was handled by the logical empiricists. They used two main approaches. One was to formulate an inductive logic that looked as much as possible like deductive logic, borrowing ideas from deductive logic whenever possible. That was Carl Hempel's approach. The other approach, used by Rudolf Carnap, was to apply the mathematical theory of probability. In the next two sections of this chapter, I will discuss some famous problems for logical empiricist theories of confirmation. The problems are especially easy to discuss in the context of Hempel's approach, which was simpler than Carnap's. A detailed examination of Carnap is beyond the scope of this book. Through his career, Carnap developed very sophisticated models of confirmation using probability theory applied to artificial languages. Problems kept arising. More and more special assumptions were needed to make the results come out right. There was never a knockdown argument against him, but the project came to seem less and less relevant to real science, and it eventually ran out of steam (Howson and Urbach 1993).

Although Carnap's approach to analyzing confirmation did not work out, the idea of using probability theory to understand confirmation remains popular and has been developed in new ways. Certainly this looks like a good approach; it does seem that observing the raised iridium level in the earth's crust made the Alvarez meteor hypothesis *more probable* than before. In chapter 14 I will describe new ways to use probability theory to understand the confirmation of theories.

Before moving on to some famous puzzles, I will discuss a simple proposal that may have occurred to you.

The term *hypothetico-deductivism* is used in several ways by people writing about science. Sometimes it is used to describe a simple view about testing and confirmation. According to this view, hypotheses in science are confirmed when their logical consequences turn out to be true. This idea covers a variety of cases; the confirmation of a white-swan generalization by observing white swans is one case, and another is the confirmation of a hypothesis about an asteroid impact by observations of the true consequences of this hypothesis.

As Clark Glymour has emphasized (1980), an interesting thing about this idea is that it is hopeless when expressed in a simple way, but something like it seems to fit well with many episodes in the history of science. One problem is that a scientific hypothesis will only have consequences of a testable kind when it is combined with other assumptions, as we have seen. But put that problem aside for a moment. The suggestion above is that a theory is confirmed when a true statement about observables can be derived from it. This claim is vulnerable to many objections. For example, any theory *T* deductively implies *T-or-S*, where *S* is any sentence at all. But *T-or-S* can be conclusively established by observing the truth of *S*. Suppose *S* is observational. Then we can establish *T-or-S* by observation, and that confirms *T*. This is obviously absurd. Similarly, if theory *T* implies observation *E*, then the theory *T&S* implies *E* as well. So *T&S* is confirmed by *E*, and *S* here could be anything at all. (Note the similarity here to a problem discussed at the beginning of section 2.4.) There are many more cases like this.

The situation is strange, and some readers may feel exasperation at this point. People do often regard a scientific hypothesis as supported when its consequences turn out to be true; this is taken to be a routine and reasonable part of science. But when we try to summarize this idea using simple logic, it seems to fall apart. Does the fault lie with the original idea, with our summary of the idea using basic logic, or with basic logic itself? The logical empiricist response was to hang steadfastly onto the logic, and often to hang onto their translations of ideas about science into a logical framework as well. This led them to question or modify some very reasonable-looking ideas about evidence and testing. But it is hard to work out where the fault really lies.

A related feature of logical empiricism is the use of simplified and artificial cases rather than cases from real science. The logical empiricists sought to strip the problem of confirmation down to its bare essentials, and they saw these essentials in formal logic. But to many, philosophy of science seemed to be turning into an exercise in "logic-chopping" for its own sake. And as we will see in the next sections, even the logic-chopping did not go well.

Despite this, there is a lot to learn from the problems faced by logical empiricism. Confirmation really *is* a puzzling thing. Let us look at some famous puzzles.

3.3 The Ravens Problem

The logical empiricists put much work into analyzing the confirmation of generalizations by observations of their instances. At this point we will switch birds, in accordance with tradition. How is it that repeated observations of black ravens can confirm the generalization that all ravens are black?

First I will deal with a simple suggestion that will not work. Some readers might be thinking that if we observe a large number of black ravens and no nonblack ones, then at least we are cutting down the number of ways in which the hypothesis that all ravens are black might be wrong. As we see each raven, there is one less raven that might fail to fit the theory. So in some sense, the chance that the hypothesis is true should be slowly increasing. But this does not help much. First, the logical empiricists were concerned to deal with the case where generalizations cover an infinite number of instances. In that case, as we see each raven we are not reducing the number of ways in which the hypothesis might fail. Also, note that even if we forget this problem and consider a generalization covering just a finite number of cases, the kind of support that is analyzed here is a very weak one. That is clear from the fact that we get no help with the problem of *projection.* As we see each raven we know there is one less way for the generalization to be false, but this does not tell us anything about what to expect with the *next* raven we see.

So let us look at the problem differently. Hempel suggested that, as a matter of logic, all observations of black ravens confirm the generalization that all ravens are black. More generally, any observation of an F that is also G supports the generalization "All F's are G." He saw this as a basic fact about the logic of support.

This looks like a reasonable place to start. And here is another obvious-looking point: any evidence that confirms a hypothesis *H* also confirms any hypothesis that is logically equivalent to *H*.

What is logical equivalence? Think of it as what we have when two sentences say the same thing in different terms. More precisely, if *H* is logically equivalent to H^*, then it is impossible for H to be true but H^* false, or vice versa.

But these two innocent-looking claims generate a problem. In basic logic the hypothesis "All ravens are black" is logically equivalent to "All nonblack things are not ravens." Let us look at this new generalization. "All nonblack things are not ravens" seems to be confirmed by the observation of a white shoe. The shoe is not black, and it's not a raven, so it fits the hypothesis. But given the logical equivalence of the two hypotheses, anything that confirms one confirms the other. So the observation of a white shoe confirms the hypothesis that all ravens are black! That sounds ridiculous. As Nelson Goodman (1955) put it, we seem to have the chance to do a lot of "indoor ornithology"; we can investigate the color of ravens without ever going outside to look at one.

This simple-looking problem is hard to solve. Debate about it continues. Hempel himself was well aware of this problem—he is the one who originally thought of it. But there has not been a solution proposed that everyone (or even most people) have agreed upon.

One possible reaction is to accept the conclusion. This was Hempel's response. Observing a white shoe *does* confirm the hypothesis that all ravens are black, though presumably only by a tiny amount. Then we can keep our simple rule that whenever we have an "All F's are G" hypothesis, any observation of an F that is G confirms it and also confirms everything logically equivalent to "All F's are G." Hempel stressed that, logically speaking, an "All F's are G" statement is not a statement about F's but a statement about everything in the universe—the statement that if something is an F then it is G. We should note that according to this reply, the observation of the white shoe also confirms the hypothesis that all ravens are green, that all aardvarks are blue, and so on. Hempel was comfortable with this situation, but most others have not been.

A multitude of other solutions have been proposed. I will discuss just two ideas, which I regard as being on the right track.

Here is the first idea. Perhaps observing a white shoe or a black raven *may or may not* confirm "All ravens are black." It depends on other factors. Suppose we know, for some reason, that either (1) all ravens are black and ravens are extremely rare, or else (2) most ravens are black, a few are white, and ravens are common. Then a casual observation of a black raven will support (2), a hypothesis that says that not all ravens are black. If all ravens were black, we should not be seeing them at all. Observing a white shoe, similarly, may or may not confirm a given hypothesis, depending on what else we know. This reply was first suggested by I. J. Good (1967).

Good's move is very reasonable. We see here a connection to the issue of holism about testing, discussed in chapter 2. The relevance of an observation to a hypothesis is not a simple matter of the content of the two statements; it depends on other assumptions as well. This is so even in the simple case of a hypothesis like "All F's are G" and an observation like "Object A is both F and G." Good's point also reminds us how artificially simplified the standard logical empiricist examples are. No biologist would seriously wonder whether seeing thousands of black ravens makes it likely that all ravens are black. Our knowledge of genetics and bird coloration leads us to expect some variation, such as cases of albinism, even when we have seen thousands of black ravens and no other colors.

Here is a second suggestion about the ravens, which is consistent with Good's idea but goes further. Whether or not a black raven or a white shoe confirms "All ravens are black" might depend on the *order* in which you learn of the two properties of the object.

Suppose you hypothesize that all ravens are black, and someone comes up to you and says, "I have a raven behind my back; want to see what color it is?" You should say yes, because if the person pulls out a white raven, your theory is refuted. You need to find out what is behind his back. But suppose the person comes up and says, "I have a black object behind my back; want to see whether it's a raven?" Then it does not matter to you what is behind his back. You think that all ravens are black, but you don't have to think that all black things are ravens. In both cases, suppose the object behind his back is a black raven and he does show it to you. In the first situation, your observation of the raven seems relevant to your investigation of raven color, but in the other case it's irrelevant.

So perhaps the "All ravens are black" hypothesis is only confirmed by a black raven when this observation had the *potential to refute* the hypothesis, only when the observation was part of a genuine test.

Now we can see what to do with the white shoe. You believe that all ravens are black, and someone comes up and says, "I have a white object behind my back; want to see what it is?" You should say yes, because if he has a raven behind his back your hypothesis is refuted. He pulls out a shoe, however, so your hypothesis is OK. Then someone comes up and says, "I have a shoe behind my back; want to see what color it is?" In this case you need not care. It seems that in the first of these two cases, you have gained some support for the hypothesis that all ravens are black. In the second case you have not.

So perhaps some white-shoe observations *do* confirm "All ravens are black," and some black-raven observations *don't*. Perhaps there is only confirmation when the observations arise during a genuine test, a test that has the potential to disconfirm as well as confirm.

Hempel saw the possibility of a view like this. His responses to Good's argument and to the order-of-observation point were similar, in fact. He said he wanted to analyze a relation of confirmation that exists *just between a hypothesis and an observation itself,* regardless of extra information we might have, and regardless of the order in which observations were made. But perhaps Hempel was wrong here; there is *no such relation.* We cannot answer the question of whether an observation of a black raven confirms the generalization unless we know something about the way the observation was made and unless we make assumptions about other matters as well.

Hempel thought that some observations are just "automatically" relevant to hypotheses, regardless of what else is going on. That is true in the case of the deductive refutation of generalizations; no matter how we come to see a nonblack raven, that is bad news for the "All ravens are black" hypothesis. But what is true for deductive *dis*confirmation is not true for confirmation.

Clearly this discussion of order-of-observation does not entirely solve the ravens problem. *Why* does order matter, for example, and what if both properties are observed at once? I will return to this issue in chapter 14, using a more complex framework. Putting it briefly, we can only understand confirmation and evidence by taking into account the *procedures* involved in generating data. Or so I will argue.

I will make one more comment on the ravens problem. This one is a digression, but it does help illustrate what is going on. In psychology there is a famous experiment called the "selection task" (Wason and Johnson-Laird 1972). The experiment has been used to show that many people (including highly educated people) make bad logical errors in certain circumstances. The experimental subject is shown four cards with half of each card masked. The subject is asked to answer this question: "Which masks do you have to remove to know whether it is true that if there is a circle on the left of a card, there is a circle on the right as well?" See fig. 3.1 and try to answer the question yourself before reading the next paragraph.

Large majorities of people in many (though not all) versions of this experiment give the wrong answer. Many people tend to answer "only card A" or "card A and card C." The right answer is A and D. Compare this to the ravens problem; the problems have the same structure. I am sure Hempel would have given the right answer if he had been a subject in the four-card experiment, but the selection task might show something interesting about

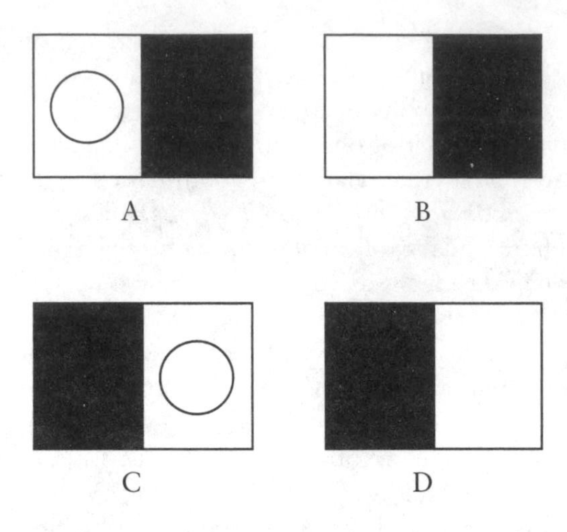

Fig. 3.1
The Wason selection task

why confirmation has been hard to analyze. For some reason it is difficult for people to see the importance of "card D" tests in cases like this, and it is easy for people to wrongly think that "card C" tests are important. If you are investigating the hypothesis that all ravens are black, card D is analogous to the situation when someone says he has a white object behind his back. Card C is analogous to the situation where he says he has a black object behind his back. Card D is a real test of the hypothesis, but card C is not. Unmasking Card C is evidentially useless, even though it may fit with what the hypothesis says. Not all observations of cases that fit a hypothesis are useful as tests.

3.4 Goodman's "New Riddle of Induction"

In this section I will describe an even more famous problem, revealed by Nelson Goodman (1955). This argument looks strange, and it is easy to misinterpret. But the issues it raises are very deep.

First we need to be clear about what Goodman was trying to do with his argument. His primary goal was to show that there cannot be a purely "formal" theory of confirmation. He does not think that confirmation is impossible, or that induction is a myth. He just thinks they work differently from the way many philosophers—especially logical empiricists—have thought.

What is a "formal" theory of confirmation? The easiest way to explain this is to look at deductive arguments. Recall the most famous deductively valid argument:

Argument 1

PREMISES All men are mortal.
Socrates is a man.

CONCLUSION Socrates is mortal.

The premises, if they are true, guarantee the truth of the conclusion. But the fact that the argument is a good one does not have anything in particular to do with Socrates or manhood. Any argument that has the same form is just as good. That form is as follows:

All F's are G.
a is an F.

---

a is G.

Any argument with this form is deductively valid, no matter what we substitute for "F," "G," and "a." As long as the terms we substitute pick out definite properties or classes of objects, and as long as the terms retain the same meaning all the way through the argument, the argument will be valid.

So the deductive validity of arguments depends only on the form or pattern of the argument, not the content. This is one of the features of deductive logic that the logical empiricists wanted to build into their theory of induction and confirmation. Goodman aimed to show that this is impossible; there can never be a formal theory of induction and confirmation.

How did Goodman do it? Consider argument 2.

Argument 2
All the many emeralds observed, in diverse circumstances,
prior to 2010 A.D. have been green.

===

All emeralds are green.

This looks like a good inductive argument. (Like some of the logical empiricists, I use a double line between premises and conclusion to indicate that the argument is not supposed to be deductively valid.) The argument does not give us a guarantee; inductions never do. And if you would prefer to express the conclusion as "probably, all emeralds are green" that will not make any difference to the rest of the discussion.

(If you know something about minerals, you might object that emeralds are regarded as green by definition: emeralds are beryl crystals made green by trace amounts of chromium. Please just regard this as another unfortunate choice of example by the literature.)

Now consider argument 3:

Argument 3
All the many emeralds observed, in diverse circumstances,
prior to 2010 A.D. have been grue.

═══════════════════════════════

All emeralds are grue.

Argument 3 uses a new word, "grue." We define "grue" as follows:

GRUE: An object is *grue* if and only if it was first observed before 2010 A.D. and is green, *or* if it was not first observed before 2010 A.D. and is blue.

The world contains lots of grue things; there is nothing strange about grue objects, even though there is something strange about the word. The grass outside my door as I write this is grue. The sky outside on July 1, 2020, will be grue, if it is a clear day. An individual object does *not* have to change color in order to be grue—this is a common misinterpretation. Anything green that has been observed before 2010 passes the test for being grue. So, all the emeralds we have seen so far have been grue.

Argument 3 does *not* look like a good inductive argument. Argument 3 leads us to believe that emeralds observed in the future will be blue, on the basis of previously observed emeralds being green. The argument also conflicts with argument 2, which looks like a good argument. But arguments 2 and 3 have *exactly the same form*. That form is as follows:

All the many E's observed, in diverse circumstances,
prior to 2010 A.D., have been G.

═══════════════════════════════

All E's are G.

We could represent the form even more schematically than this, but that does not matter to the point. Goodman's point is that two inductive arguments can have the exact same form, but one argument can be good while the other is bad. So what makes an inductive argument a good or bad one cannot be just its form. Consequently, there can be no purely formal theory of induction and confirmation. Note that the word "grue" works perfectly well in *de*ductive arguments. You can use it in the form of argument 1, and it will cause no problems. But induction is different.

Suppose Goodman is right, and we abandon the idea of a formal theory of induction. This does not end the issue. We still need to work out *what* exactly is wrong with argument 3. This is the new riddle of induction.

The obvious thing to say is that there is something wrong with the word "grue" that makes it inappropriate for use in inductions. So a good theory of induction should include a *restriction* on the terms that occur in inductive arguments. "Green" is OK and "grue" is not.

This has been the most common response to the problem. But as Goodman says, it is very hard to spell out the details of such a restriction. Suppose we say that the problem with "grue" is that its definition includes a reference to a specific time. Goodman's reply is that whether or not a term is defined in this way depends on which language we take as our starting point. To see this, let us define a new term, "bleen."

BLEEN: An object is *bleen* if and only if it was first observed before 2010 A.D. and is blue, *or* if it was not first observed before 2010 A.D. and is green.

We can use the English words "green" and "blue" to define "grue" and "bleen," and if we do so we must build a reference to time into the definitions. But suppose we spoke a language that was like English except that "grue" and "bleen" were basic, familiar terms and "green" and "blue" were not. Then if we wanted to define "green" and "blue," we would need a reference to time.

GREEN: An object is *green* if and only if it was first observed before 2010 A.D. and is grue, *or* if it was not first observed before 2010 A.D. and is bleen.

(You can see how it will work for "blue.") So Goodman claimed that whether or not a term "contains a reference to time" or "is defined in terms of time" is a *language-relative* matter. Terms that look OK from the standpoint of one language will look odd from another. So if we want to rule out "grue" from inductions because of its reference to time, then whether an induction is good or bad will depend on *what language we treat as our starting point.* Goodman thought this conclusion was fine. A good induction, for Goodman, must use terms that have a history of normal use in our community. That was his own solution to his problem. Most other philosophers did not like this at all. It seemed to say that the value of inductive arguments depended on irrelevant facts about which language we happen to use.

Consequently, many philosophers have tried to focus not on the words "green" and "grue" but on the *properties* that these words pick out, or the *classes* or *kinds* of objects that are grouped by these words. We might argue that greenness is a natural and objective feature of the world, and grueness is not. Putting it another way, the green objects make up a "natural kind," a kind unified by real similarity, while the grue objects are an artificial or arbitrary collection. Then we might say: a good induction has to use terms that we have reason to believe pick out natural kinds. Taking this approach plunges us into hard problems in other parts of philosophy. What *is* a property? What *is* a "natural kind"? These are problems that have been controversial since the time of Plato.

Although Goodman's problem is abstract, it has interesting links to real problems in science. In fact, Goodman's problem encapsulates within it several distinct hard methodological issues in science; that is partly why the problem is so interesting. First, there is a connection between Goodman's problem and the "curve-fitting problem" in data analysis. Suppose you have a set of data points in the form of x and y values, and you want to discern a general relationship expressed by the points by fitting a function to them. The points in figure 3.2 fall almost exactly on a straight line, and that seems to give us a natural prediction for the y value we expect for $x = 4$. However, there is an infinite number of different mathematical functions that fit our three data points (as well or better) but which make different predictions for the case of $x = 4$. How do we know which function to use? Fitting a strange function to the points seems to be like preferring a grue induction over a green induction when inferring from the emeralds we have seen.

Scientists dealing with a curve-fitting problem like this may have extra information telling them what sort of function is likely here, or they may prefer a straight line on the basis of *simplicity*. That suggests a way in which we might deal with Goodman's original problem. Perhaps the green induction is to be preferred on the basis of its simplicity?

That might work, but there are problems. First, is it really so clear that the green induction is simpler? Goodman will argue that the simplicity of an inductive argument depends on which language we assume as our starting point, for the kinds of reasons given earlier in this section. For Goodman, what counts as a simple pattern depends on which language you speak or which categorization you assume. Also, though a preference for

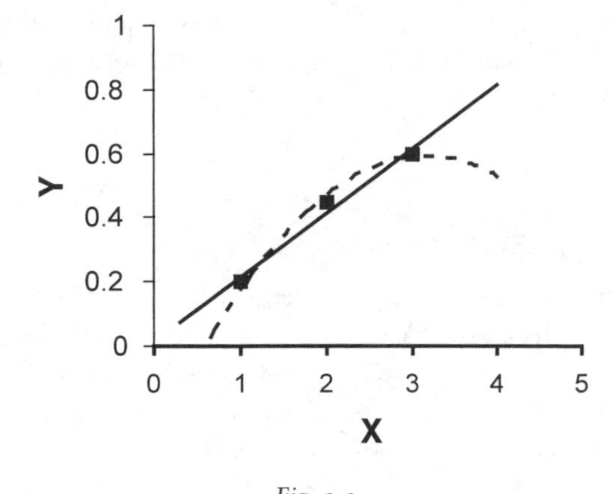

Fig. 3.2
The curve-fitting problem

simplicity is very common in science, such a preference is often hard to justify. Simpler theories are easier for us to work with, but that does not seem to give us reason to prefer them if we are seeking to learn what the world is really like. Why should the world be simple rather than complex?

Earlier I mentioned attempts to solve Goodman's problem using the idea of a "natural kind," a collection unified by real similarity as opposed to stipulation or convention. Though this term is philosophical, a lot of argument within science is concerned just this sort of problem—with getting the right *categories* for prediction and extrapolation. The problem is especially acute in sciences like economics and psychology that deal with complex networks of similarities and differences across the cases they try to generalize about. Do all economies with very high inflation fall into a natural kind that can be used to make general predictions? Are the mental disorders categorized in psychiatric reference books like the *DSM IV* really natural kinds, or have we applied standard labels like "schizophrenia" to groups of cases that have no real underlying similarity? The periodic table of elements in chemistry seems to pick out a set of real natural kinds, but is this something we can hope for in all sciences? If so, what does that tell us about inductive arguments in different fields?

That concludes our initial foray into the problems of induction and confirmation. These problems are simple, but they are very resistant to solution. For a good part of the twentieth century, it seemed that even the most innocent-looking principles about induction and confirmation led straight into trouble.

Later (especially in chapter 14) I will return to these problems. But in the next chapter we will look at a philosophy that gets a good part of its motivation from the frustrations discussed in this chapter.

Further Reading

Once again, Hempel's *Aspects of Scientific Explanation* (1965) is a key source, containing a long (and exhausting) chapter on confirmation. Skyrms, *Choice and Chance* (2000), is a classic introductory book on these issues, and it introduces probability theory as well. Even though it argues for a view that will not be discussed until chapter 14, Howson and Urbach's *Scientific Reasoning* (1993) is a useful introduction to various approaches to confirmation. It has the most helpful short summary of Carnap's ideas that I have read. Carnap's magnum opus on these issues is his *Logical Foundations of Probability* (1950). For a discussion of explanatory inference, see Lipton, *Inference to the Best Explanation* (1991).

For the use of order-of-observation to address the ravens problem, see Horwich, *Probability and Evidence* (1982), but you should probably read chapter 14 of this book first.

Goodman's most famous presentation of his "new riddle of induction" is in *Fact, Fiction & Forecast* (1955). The problem is in chapter 3 (along with other interesting ideas), and his solution is in chapter 4. His subsequent papers on the topic are collected in *Problems and Projects* (1972). Douglas Stalker has edited a collection on Goodman's riddle, called *Grue!* (1994). It includes a very detailed bibliography. The Quine and Jackson papers are particularly good.

For discussions of properties and kinds, and their relevance to induction, see Armstrong 1989, Lewis 1983, Dupre 1993, and Kornblith 1993. (These are fairly advanced discussions, except for Armstrong's, which is introductory.) There is a good discussion of simplicity in Sober 1988.

4

Popper: Conjecture and Refutation

4.1 Popper's Unique Place in the Philosophy of Science

Karl Popper is the only philosopher discussed in this book who is regarded as a hero by many scientists. Attitudes toward philosophy among scientists vary, but hardly ever does a philosopher succeed in *inspiring* scientists in the way Popper has. It is also rare for a philosopher's view of science to be used within a scientific debate to justify one position over another. This has happened with Popper too. Within biology, recent debates about the classification of organisms and about ecology have both seen Popper's ideas used in this way (Hull 1999). I once went to a lecture by a famous virologist who had won a Nobel Prize in medicine, to hear about his work. What I heard was mostly a lecture about Popper. In 1965, Karl Popper even became *Sir* Karl Popper, knighted by the queen of England.

Popper's appeal is not surprising. His view of science is centered around a couple of simple, clear, and striking ideas. His vision of the scientific enterprise is a noble and heroic one. Popper's theory of science has been criticized a great deal by philosophers over the years. I agree with many of these criticisms and don't see any way for Popper to escape their force. Despite the criticism, Popper's views continue to have an important place in philosophy and continue to appeal to many working scientists.

4.2 Popper's Theory of Science

Popper began his intellectual career in Vienna, between the two world wars. He was not part of the Vienna Circle, but he did have contact with the logical positivists. This contact included a lot of disagreement, as Popper developed his own distinctive position. Popper does count as an "empiricist" in the broad sense used in this book, but he spent a lot of time distinguishing his views from more familiar versions of empiricism. Like the logical positivists, Popper left Europe upon the rise of Nazism, and after spending the war years in New Zealand, he moved to the London School of Economics, where he remained for the rest of his career. There he built up a loyal group of allies, whom he often accused of disloyalty. His seminar series at the London School of Economics became famous for its grueling questioning and for the fact that speakers had a difficult time actually presenting much of their lectures, because of Popper's interruptions.

Popper once had a famous confrontation with Wittgenstein, on the latter's turf at Cambridge University. One version of the story, told by Popper himself, has Wittgenstein brandishing a fireplace poker during a discussion of ethical rules, leading Popper to give as an example of an ethical rule: "not to threaten visiting lecturers with pokers." Wittgenstein stormed out. Other versions of the story, including those told by Wittgenstein's allies, deny Popper's account (see Edmonds and Eidinow 2001 for this controversy).

The logical positivists developed their theory of science as part of a general theory of language, meaning, and knowledge. Popper was not much interested in these broader topics, at least initially; his primary aim was to understand science. As his first order of business, he wanted to understand the difference between scientific theories and nonscientific theories. In particular, he wanted to distinguish science from "pseudo-science." Unlike the logical positivists, he did not regard pseudo-scientific ideas as meaningless; they just weren't science. For Popper, an inspiring example of genuine science was the work of Einstein. Examples of pseudo-science were Freudian psychology and Marxist views about society and history.

Popper called the problem of distinguishing science from non-science the "problem of demarcation." All of Popper's philosophy starts from his proposed solution to this problem. "Falsificationism" was the name Popper gave to his solution. Falsificationism claims that *a hypothesis is scientific if and only if it has the potential to be refuted by some possible observation.* To be scientific, a hypothesis has to take a risk, has to "stick its neck out." If a theory takes no risks at all, because it is compatible with every possible observation, then it is not scientific. As I said above, Popper held that Marx's and Freud's theories were not scientific in this sense. No matter what happens, Popper thought, a Marxist or a Freudian can fit it somehow into his theory. So these theories are never exposed to any risks.

So far I have described Popper's use of falsifiability to distinguish scientific from nonscientific theories. Popper also made use of the idea of falsification in a more far-reaching way. He claimed that *all* testing in science has the form of attempting to refute theories by means of observation. And crucially, for Popper it is never possible to confirm or establish a theory by showing its agreement with observations. *Confirmation is a myth.* The only thing an observational test can do is to show that a theory is false. So the truth of a scientific theory can never be supported by observational evidence, not even a little bit, and not even if the theory makes a huge number of predictions that all come out as expected.

As you might think, Popper was a severe critic of the logical empiricists' attempts to develop a theory of confirmation or "inductive logic." The problems they encountered, some of which I discussed in chapter 3, were music to his ears. Popper, like Hume, was an inductive skeptic, and Popper was skeptical about all forms of confirmation and support other than deductive logic itself.

Skepticism about induction and confirmation is a much more controversial position than Popper's use of falsification to solve the demarcation problem. Most philosophers of science have thought that if induction and confirmation are just myths, that is very bad news for science. Popper tried to argue that there is no reason to worry; induction is a myth, but science does not need it anyway. So inductive skepticism, for Popper, is no threat to the rationality of science. In the opinion of most philosophers, Popper's attempt to defend this radical claim was not successful, and some of his discussions of this topic are rather misleading to readers. As a result, some of the scientists who regard Popper as a hero do not realize that Popper believed it is never possible to confirm a theory, not even slightly, and no matter how many observations the theory predicts successfully.

Popper placed great emphasis on the idea that we can never be *completely sure* that a theory is true. After all, Newton's physics was viewed as the best-supported theory ever, but early in the twentieth century it was shown to be false in several respects. However, almost all philosophers of science accept that we can never be 100 percent certain about factual matters, especially those discussed in science. This position, that we can never be completely certain about factual issues, is often known as *fallibilism* (a term due to C. S. Peirce). Most philosophers of science accept fallibilism. The harder question is whether or not we can be reasonable in *increasing* our confidence in the truth of a theory when it passes observational tests. Popper said no. The logical empiricists and most other philosophers of science say yes.

So Popper had a fairly simple view of how testing in science proceeds. We take a theory that someone has proposed, and we deduce an observational prediction from it. We then check to see if the prediction comes out as the theory says it will. If the prediction fails, then we have refuted, or falsified, the theory. If the prediction does come out as predicted, then all we should say is that *we have not yet falsified the theory.* For Popper, we cannot conclude that the theory is true, or that it is probably true, or even that it is more likely to be true than it was before the test. The theory *might* be true, but we can't say more than that.

We then try to falsify the theory in some other way, with a new prediction. We keep doing this until we have succeeded in falsifying it. What if years pass and we seem to never be able to falsify a theory, despite repeated tests? We can say that the theory has now survived repeated attempts to falsify it, but that's all. We never increase our confidence in the truth of the theory; and ideally, we should never stop trying to falsify it. That's not to say we should spend all our time testing theories that have passed tests over and over again. We do not have the time and resources to test everything that could be tested. But that is just a practical constraint. According to Popper, we should always retain a *tentative* attitude toward our theories, no matter how successful they have been in the past.

In defending this view, Popper placed great emphasis on the difference between confirming and disconfirming statements of scientific law. If someone proposes a law of the form "All F's are G," all it takes is one observation of an F that is not a G to falsify the hypothesis. This is a matter of deductive logic. But it is never possible to assemble enough observations to conclusively demonstrate the truth of such a hypothesis. You might wonder about situations where there is only a small number of F's and we could hope to check them all. Popper and the logical empiricists regarded these as unimportant situations that do not often arise in science. Their aim was to describe testing in situations where there is a huge or infinite number of cases covered by a hypothesized law or generalization. So Popper stressed that universal statements are hard or impossible to verify but easy, in principle, to falsify. The logical empiricist might reply that statements of the form "Some F's are G" have the opposite feature; they are easy to verify but hard or impossible to falsify. But Popper claimed (and the logical empiricists tended to agree) that real scientific theories rarely take this form, even though some statements in science do.

Despite insisting that we can never support or confirm scientific theories, Popper believed that science is a search for true descriptions of the world. How can one search for truth if confirmation is impossible?

This is an unusual kind of search. We might compare it to a certain kind of search for the Holy Grail, conducted by an imaginary medieval knight. Suppose there are lots of grails around, but only one of them is holy. In fact, the number of nonholy grails is infinite or enormous, and you will never encounter them all in a lifetime. All the grails glow, but only the Holy Grail glows forever. The others eventually stop glowing, but there is no telling when any particular nonholy grail will stop glowing. All you can do is pick up one grail and carry it around and see if it keeps on glowing. You are only able to carry one at a time. If the one you are carrying is the Holy Grail, it will never stop glowing. But you would never *know* if you currently had the Holy Grail, because the grail you are carrying might stop glowing at any moment. All you can do is reject grails that are clearly not holy (since they stop glowing at some point) and keep picking up a new one. You will eventually die (with no afterlife, in this scenario) without knowing whether you succeeded.

This is similar to Popper's picture of science's search for truth. All we can do is try out one theory after another. A theory that we have failed to falsify up till now *might,* in fact, be true. But if so, we will never know this or even have reason to increase our confidence.

4.3 Popper on Scientific Change

So far I have described Popper's views about the demarcation of science from non-science and the nature of scientific testing. Popper also used the idea of falsification to propose a theory of scientific *change.*

Popper's theory has an appealing simplicity. Science changes via a two-step cycle that repeats endlessly. Stage 1 in the cycle is *conjecture*—a scientist will offer a hypothesis that might describe and explain some part of the world. A good conjecture is a *bold* one, one that takes a lot of risks by making novel predictions. Stage 2 in the cycle is *attempted refutation*—the hypothesis is subjected to critical testing, in an attempt to show that it is false. Once the hypothesis is refuted, we go back to stage 1 again—a new conjecture is offered. That is followed by stage 2, and so on.

As the process moves along, it is natural for a scientist to propose conjectures that have some relation to previous ones. A theoretical idea can be refined and modified via many rounds of conjecture and refutation. That is fine, for Popper, though it is not essential. One thing that a scientist should *not* do, however, is to react to the falsification of one conjecture by cooking up a new conjecture that is designed to just avoid the problems revealed by earlier testing, and which goes no further. We should not make *ad hoc* moves that merely patch the problems found in earlier conjectures. Instead, a scientist should constantly strive to *increase* the breadth of application of a theory and increase the precision of its predictions. That means constantly trying to increase the "boldness" of conjectures.

What sort of theory is this? Popper intended it as a description of the general pattern that we *actually* see in science, and as a description of *good* scientific behavior as well. He accepted that not all scientists succeed in sticking to this pattern of behavior all the time. Sometimes people become too wedded to their hypotheses; they refuse to give them up when testing tells them to. But Popper thought that a lot of actual scientific behavior does follow this pattern and that we see it especially in great scientists such as Einstein. For Popper, a good or great scientist is someone who combines two features, one corresponding to each stage of the cycle. The first feature is an ability to come up with imaginative, creative, and risky ideas. The second is a hard-headed willingness to subject these imaginative ideas to rigorous critical testing. A good scientist has a creative, almost artistic, streak and a tough-minded, no-nonsense streak. Imagine a hard-headed cowboy out on the range, with a Stradivarius violin in his saddlebags. (Perhaps at this point you can see some of the reasons for Popper's popularity among scientists.)

Popper's view here can apparently be applied in the same way to *individuals* and to *groups* of scientists. An isolated individual can behave scientifically by engaging in the process of conjecture and refutation. And a collection of scientists can each, at an individual level, follow Popper's two-step procedure. But another possibility is a division of labor; one individual (or team) comes up with a conjecture, and another does the attempted refutation. Popper's basic description of the two-step conjecture-and-refutation pattern of science seems compatible with all these possibilities. But the case where individual A does the conjecture and individual B does the refutation will be suspicious to Popper. If individual A is a true scientist, he should take a critical attitude toward his own ideas. If individual A is completely fixated on his conjecture, and individual B is fixated on showing that A is wrong in order to advance his *own* conjecture, this is not good scientific behavior according to Popper.

This raises an interesting question. Empiricist philosophies stress the virtues of open-mindedness, and Popper's view is no exception. But perhaps an open-minded *community* can be built out of a collection of rather closed-minded *individuals*. If actual scientists are wedded to their own conjectures, but each is wedded to a *different* conjecture and would like to prove the others wrong, shouldn't the overall process of conjecture and refutation work? What is wrong with the situation where B's role is to critically test A's ideas? So long as the testing occurs, what does it matter whether A or B does it? One problem is that if everyone is so closed-minded, the results of the test might have no impact on what people believe. Perhaps the young and tender minds of incoming graduate students could be the community's source of flexibility; unsuccessful theories will attract no new recruits and will die with their originators. This would be a rather slow way for science to change (but many would argue that we do see cases like this).

In later chapters of this book, we will look at theories that focus on social structure in science, and at various kinds of division of labor between individual scientists. Although Popper did stress community standards in science, he did seem to have a picture in which the good scientist should, *as an individual,* have the willingness to perform both the imaginative and the critical roles. A good scientist should retain a tentative attitude toward all theories, including his own.

I will make one more point before moving on to criticisms of Popper. The two-step process of conjecture and refutation that Popper describes has a striking resemblance to another two-step process: Darwin's explanation of biological evolution in terms of *variation and natural selection.* In science according to Popper, scientists toss out conjectures that are subjected to critical testing. In evolution, according to both Darwin himself and more recent versions of evolutionary theory, populations evolve via a process in which variations appear in organisms in a random or "undirected" way, and these novel characteristics are "tested" through their effects on the organism in its interactions with the environment. Variations that help organisms to survive and reproduce, and which are of the kind that gets passed on in reproduction, tend to be preserved and become more common in the population over time.

Ironically, at one time Popper thought that Darwinism is not a scientific theory, but he later retracted that claim. In any case, both Popper and others have explored the analogy between Popperian science and Darwinian evolution in detail. The analogy should not be taken *too* seriously; evolution is not a process in which populations really "search" for anything, in the way that scientists search for good theories, and there are other crucial differences too. But the similarity is certainly interesting. Analogies between science and evolution will come again in later chapters (6 and 11).

4.4 Objections to Popper on Falsification

Let us now turn to a critical assessment of Popper's ideas. We should start with his solution to the demarcation problem. Is falsifiability a good way to distinguish scientific ideas from nonscientific ones?

Let me first say that I think this question probably has no answer in the form in which Popper expressed it. We should not expect to be able to go through a list of statements or theories and label them "scientific" or "not scientific." However, I suggest that something fairly similar to Popper's question about demarcation does make sense: can we describe a distinctive scientific *strategy* of investigating the world, a scientific way of *handling* ideas?

Some of Popper's ideas are useful in trying to answer this question. In particular, Popper's claim that scientific theories should take *risks* is a good one; this will be followed up in the last section of this chapter. But Popper had an overly simple picture of *how* this risk-taking works.

For Popper, theories have the form of generalizations, and they take risks by prohibiting certain kinds of particular events from being observed. If we believe that all pieces of iron, of whatever size and shape, expand when heated, then our theory forbids the observation of something that we know to be a piece of iron contracting when heated. A problem may have occurred to you: how sure can we be that, if we see a piece of "iron" contracting when heated, that it is really iron? We might also have doubts about our measurements of the contraction and the temperature change. Maybe the generalization about iron expanding when heated is true, but our assumptions about the testing situation and our ability to know that a sample is made of iron are false.

This problem is a reappearance of an issue discussed in chapter 2: holism about testing. Whenever we try to test a theory by comparing it with observations, we must make a large number of additional assumptions in order to bring the theory and the observations into "contact" with each other. If we want to test whether iron always expands when heated, we need to make assumptions about our ability to find or make reasonably pure samples of iron. If we want to test whether the amounts of the bases C and G are equal and the amounts of A and T are equal in all samples of DNA (Chargaff's rules), we need to make a lot of assumptions about our chemical techniques. If we observe an unexpected result (iron contracting on heating, twice as much C as G in a sample of DNA), it is always possible to blame one of these extra assumptions rather than the theory we are trying to test. In extreme cases, we might even claim that the apparent observation was completely misunderstood or wrongly described by the observers. Indeed, this is not so uncommon in our attempts to work out what to make of reports of miracles and UFO abductions. So how can we really use observations to falsify theories in the way Popper wants?

This is a problem not just for Popper's solution to the demarcation problem, but for his whole theory of science as well.

Popper was well aware of this problem, and he struggled with it. He regarded the extra assumptions needed to connect theories with testing situations as scientific claims that might well be false—these are conjectures too. We can try to test these conjectures separately. But Popper conceded that logic itself can never *force* a scientist to give up a particular theory, in the face of surprising observations. Logically, it is always possible to blame other assumptions involved in the test. Popper thought that a good scientist would not try to do this; a good scientist is someone who wants to expose the theory itself to tests and will not try to deflect the blame.

Does this answer the holist objection? What Popper has done is to move from describing a characteristic of scientific *theories* to describing a characteristic of scientific *behavior.* In some ways this is a retraction of his initial aim, which was to describe something about scientific theories themselves that makes them special. That is a problem. Then again, this shift to describing scientific modes of thought and behavior, rather than theories, may well be a step forward. This will be discussed in more detail in section 4.6.

Popper also accepted that we cannot be completely certain about the observation reports that we use to falsify theories. We have to regard the acceptance of an observation report as a "decision," one that is freely made. Once we have made the decision, we can use the observation report to falsify any theory that conflicts with it. But for Popper, any falsification process is based, in the end, on a decision that could be challenged. Someone might come along later and try to show, via more testing, that the observation report was not a good one—that person might investigate whether the conditions of observation were misleading. That testing has the same conjecture-and-refutation form described earlier. So this investigation into the controversial observation ultimately depends on "decisions" too.

Is this bad news for Popper? Popper insisted that making these decisions about single observations is very different from making free decisions directly about the theories themselves. But what *sort* of difference is this? If observation reports rest on nothing more than "decisions," and these determine our choice of theories, how is that better than directly choosing the theories themselves, without worrying about observation? Or why couldn't we just "decide" to hang onto a theory and reject the observation reports that conflict with it? I am not saying that we *should* do these things, just that Popper has not given us a good reason not to do them. I believe that we should not do these things because we have good reason to believe that observation is a generally reliable way of forming beliefs. As I will argue in chapter 10, we need to make use of a scientific theory of perception at this point in the story. But that argument will have to come later. Popper himself does not try to answer these questions by giving an argument about the reliability of perception.

This point about the role of decisions affects Popper's ideas about demarcation as well as his ideas about testing. Any system of hypotheses can be held onto despite apparent falsification, if people are willing to make certain decisions. Does that mean that Popper's theory fails to differentiate between science and pseudo-science after all? The answer is "yes and no." The yes comes from the fact that scientific theories can be handled in a way that makes them immune to falsification, and nonscientific theories can be rejected if people decide to accept claims about particular matters that are incompatible with the theory. But there is a "no" part in the answer as well. A scientific theory is falsifiable via a certain *kind* of decision—a decision about an observation report. A pseudo-scientific theory, Popper says, does not clash with any possible observations. So if a pseudo-scientific theory is to be rejected, some different kind of decision must be made. We can accept, with Popper, that this is a significant difference. But Popper has not told us why this way of doing things, the scientific way, is more rational than some other way.

I have been fairly tough on Popper's views about falsification in this section, and there is another problem to discuss as well. The problem is bad for Popper, but I should emphasize that it is bad for many others as well.

What can Popper say about theories that do not claim that some observation O is forbidden, but only that it is very *unlikely*? If I believe that a certain coin is "fair," I can deduce from this hypothesis various claims about the probabilities of long "all heads" or "all tails" sequences of tosses. Suppose I observe 100 tosses turning up heads 100 times. This is very unlikely according to my hypothesis about the coin, but it is not impossible. *Any* finite stretch of heads tosses is possible with a fair coin, although longer and longer runs of heads are treated by the theory as more and more unlikely. But if a hypothesis does not *forbid* any particular observations, then, according to Popper, it is taking no risks. That seems to entail that theories that ascribe low probabilities to specific observations, but do not rule them out altogether, are unfalsifiable and hence unscientific for Popper.

Popper's response was to accept that, logically speaking, all hypotheses of this kind *are* unscientific. But this seems to make a mockery of the important role of probability in science. So Popper said that a scientist can decide that if a theory claims that a particular observation is extremely improbable, the theory *in practice* rules out that observation. So if the observation is made, the theory is, in practice, falsified. According to Popper, it is up to scientists to work out, for their own fields, what sort of probability is so low that events of that kind are treated as prohibited. So probabilistic theories can only be construed as falsifiable in a special "in practice" sense. And we have here another role for "decisions" in Popper's philosophy of science, as opposed to the constraints of logic.

Popper is right that scientists reject theories when observations occur which the theory says are highly improbable (although it is a complicated matter which *kinds* of improbability have this importance). And Popper is right that scientists spend a good deal of time working out "how improbable is *too* improbable." Complex statistical methods are used to help scientists with these decisions. But in making this move, Popper has badly damaged his original picture of science. This was a picture in which observations, once accepted, have the power to decisively refute theoretical hypotheses. That is a matter of deductive logic, as Popper endlessly stressed. Now Popper is saying that falsification can occur without its being backed up by a deductive logical relation between observation and theory.

4.5 Objections to Popper on Confirmation

As described earlier, Popper believed that theories are never confirmed by observations, and he thought inductive arguments are never justified. Popper thought that a theory of the rational choice of theories could be given entirely in terms of falsification, so he thought that rejecting induction and confirmation was no problem.

In the previous section I discussed problems with Popper's views about falsification. But let us leave those problems aside now, and assume in this section that we can use Popperian falsification as a method for decisively rejecting theories. If we make this assumption, is Popper's attempt to describe rational theory choice successful? No, it is not.

Here is simple problem that Popper has a very difficult time with. Suppose we are trying to build a bridge, and we need to use physical theories to tell us which designs are stable and will support the weight that the bridge must carry. This is a situation where we must apply our scientific theories to a practical task. As a matter of fact, engineers and scientists in this situation will undoubtedly tend to use physical theories that have survived empirical testing; they will use "tried and true" methods as far as possible. The empiricist approach to the philosophy of science holds that such a policy is rational. The problem for an empiricist philosophy is to explain in more detail *why* this policy is the right one. That task is hard, as I hope became clear in chapter 3. But let us focus on Popper, who wants to avoid the need for a theory of confirmation. How does Popper's philosophy treat the bridge-building situation?

Popper can say why we should prefer to use a theory that has not been falsified over a theory that has been falsified. Theories that have been falsified have been shown to be false (here again I ignore the problems discussed in the previous section). But suppose we have to choose between (1) a theory that has been tested many times and has passed every test, and (2) a brand new theory that has just been conjectured and has never been tested. Neither theory has been falsified. We would ordinarily think that the rational thing to do is to choose the theory that has survived testing. But what can Popper say about this choice? Why exactly would it be irrational, for Popper, to build the bridge using a brand new theory that has never been tested?

Popper recognized and struggled with this problem too. Perhaps this has been the most common objection to Popper from other empiricist philosophers (e.g., Salmon 1981). Popper is not able to give a very good reply.

Popper refuses to say that when a theory passes tests, we have more reason to believe that the theory is true. Both the untested theory and the well-tested theory are just conjectures. But Popper did devise a special concept to use in this situation. Popper said that a theory that has survived many attempts to falsify it is "corroborated." And when we face choices like the bridge-building one, it is rational to choose corroborated theories over theories that are not corroborated.

What is "corroboration"? Popper gave a technical definition and held that we can measure the amount of corroboration that a theory has at a particular time. The technicalities do not matter, though. We need to ask, What *sort* of property is corroboration? Has Popper just given a new name to confirmation? If so, he can answer the question about building the bridge, but he has given up one of his main differences from the logical empiricists and everyone else. If corroboration is totally different from confirmation—so different that we cannot regard corroboration as any guide to *truth*—then why should we choose a corroborated theory when we build the bridge? This issue has been much discussed (see Newton-Smith 1981). Popper's concept of corroboration can be interpreted in a way that makes it different from confirmation, but Popper can give no good answer to why we should choose corroborated theories over new ones when building bridges.

To understand corroboration, think of the difference between an academic transcript and a letter of recommendation. This distinction should be vivid to students! An academic transcript says what you have *done*. It measures your past performance, but it does not contain explicit predictions about what you will do in the future. A letter of recommendation usually says something about what you have done, and it also makes claims about how you are likely to do in the future. Confirmation, as understood by the logical empiricists, is something like a letter of recommendation for a scientific theory. Corroboration, for Popper, is only like an academic transcript. And Popper thought that no good reasons could be given for believing that past performance is a reliable guide to the future. So corroboration is entirely "backward-looking." Consequently, no reason can be given for building a bridge with a corroborated theory rather than a noncorroborated but unfalsified one.

I think the best thing for Popper to say about the bridge-building situation is to stick to his inductive skepticism. He should argue that we really *don't* know what will happen if we build another bridge with a design that has worked in the past. Maybe it will stay up and maybe it won't. There might also be practical reasons for choosing that design if we are very familiar with it. But if someone comes along with a brand new untested design, we won't know whether it's a bad design until we try it.

Popper liked to say here that there is no alternative policy that is *more* rational than using the familiar and well-tested design, and we do have to make *some* decision. So we can go ahead and use the established design. But as Wesley Salmon (1981) replied, this does not help at all. If confirmation does not exist, then it seems there is also no policy that is *more* rational than choosing the *un*tested design. All we have here is a kind of "tie" between the options.

For most people, this is an unsatisfactory place for a philosophy of science to end up. Inductive skepticism of this kind is hard to take seriously outside of abstract, academic discussion. However, the efforts of the last two hundred years have shown how extremely hard it is to produce a good theory of induction and confirmation. One of the valuable roles of Popper's philosophy is to show what sort of theory of science might be possible if we give up on induction and confirmation.

In the first chapter of this book, I said that few philosophers still try to give descriptions of a definite "scientific method," where this is construed as something like a recipe for science. Popper is something of an exception here, since he does come close to giving a kind of recipe (although Popper insists there is no recipe for coming up with interesting conjectures). His view has an interesting relationship to descriptions of scientific method given in science textbooks.

In many textbooks, one finds something called the "hypothetico-deductive method." Back in chapter 3, I discussed a view about confirmation that is often called "hypothetico-deductivism." Now we are dealing with a method rather than a theory of confirmation. Science textbooks are more cautious about laying out recipes for science than they used to be, but descriptions of the hypothetico-deductive method are still fairly common. Formulations of the method vary, but some are basically a combination of Popper's view of testing and a less skeptical view about confirmation. In these versions, the hypothetico-deductive method is a process in which scientists come up with conjectures and then deduce observational predictions from those conjectures. If the predictions come out as the theory says, then the theory is supported. If the predictions do not come out as the theory says, the theory is not supported and should be rejected.

This process has the basic pattern that Popper described, but the idea that theories can be "supported" by observations is *not* a Popperian idea.

The term "support" is vague, but I think discussions of the hypothetico-deductive method generally assume that if a theory makes a lot of successful predictions, we have more reason to believe that the theory is true than we had before the successful predictions were made. We will never be completely sure, of course. But the more tests a theory passes, the more confidence we can have in its truth. The idea that we can gradually increase our confidence that a theory is true is an idea that Popper rejected. As I said at the start of this chapter, some of Popper's scientific admirers do not realize that Popper's view has this feature, because some of Popper's discussions were misleading.

Other formulations of the hypothetico-deductive method include a first stage in which observations are collected and a conjecture is generated *from* these observations. Popper disagreed with this picture of scientific procedure because he argued that fact-gathering can only take place in a way guided by a conjecture. But this is a fairly minor point.

Another term that some textbooks use in discussing scientific method (though not so much any more) is "strong inference." This term was introduced by a chemist named John Platt (1964). Strong inference is roughly a Popperian kind of testing plus another further assumption, which Popper rejected. This assumption is that we can write down *all* the possible theories that might be true in some area, and test them one by one. We find the true theory by eliminating the alternatives—it's a kind of "Sherlock Holmes" method. For Popper, this is impossible. In any real case, there will be an infinite number of competing theories. So even if we eliminate ten or one hundred possibilities, there is still the same infinite number remaining. According to Popper, all we can do is to choose one theory, test it, then choose another, and so on. We can never have confidence that we have eliminated all, or most, of the alternatives. (More recent attempts to make use of this "Sherlock Holmes" method will be discussed in chapter 14.)

I have not discussed objections to Popper's theory of scientific change (section 4.3) yet, but I will do so in the next few chapters.

What is Popper's single most important and enduring contribution to philosophy of science? I'd say it is his use of the idea of "riskiness" to describe the kind of contact that scientific theories have with observation. Popper was right to concentrate on the ideas of exposure and risk in his description of science. Science tries to formulate and handle ideas in such a way that they are exposed to falsification and modification via observation. Popper's formulation is valuable because it captures the idea that theories can *appear* to have lots of contact with observation when in fact they only have a kind of "pseudo-contact" with observation because they are exposed to no risks. This is an advance in the development of empiricist views of science. Popper's analysis of *how* this exposure works does not work too well, but the basic idea is good.

4.6 Further Comments on the Demarcation Problem

Popper is onto something when he says that scientific theories should take risks. In this section I will try to develop this idea a bit differently.

Popper was interested in distinguishing scientific theories from unscientific ones, and he wanted to use the idea of risk-taking to make the distinction. But this idea of risk-taking is better used as a way of distinguishing scientific from unscientific ways of *handling* ideas. And we should not expect a sharp distinction between the two.

The scientific way of handling an idea is to try to connect it with other ideas, to embed it in a larger conceptual structure, in a way that *exposes it to observation.* This "exposure" is not a matter of simple falsification; there are many ways in which exposure to observation can be used to modify and assess an idea. But if a hypothesis is handled in a way that keeps it apart from all the risks associated with observation, that is an unscientific handling of the idea.

So it is a mistake to try to work out whether theories like Marxism or Freudianism are themselves "scientific" or not, as Popper did. A big idea like Marxism or Freudianism will have scientific and unscientific *versions,* because the main principles of the theory can be handled scientifically or unscientifically. Scientific versions of Marxism and Freudianism are produced when the main principles are connected with other ideas in a way that exposes these principles to testing. To scientifically handle the basic principles of Marxism is to try to work out what *difference* it would make to things we can observe if the Marxist principles were true. To do this it is not necessary that we write down some single observation that, if we observe it, will lead us to definitively reject the main principles of the theory. It will remain possible that an auxiliary assumption is at fault, and there is no simple recipe for adjudicating such decisions.

To continue with Popper's examples, Marxism holds that the driving force of human history is struggle between economic classes, guided by ongoing changes in economic organization. This struggle results in a predictable sequence of political changes, leading eventually to socialism. Freudianism holds that the normal development of a child includes a series of interactions and conflicts between unconscious aspects of the child's mind, where these interactions have a lot to do with resolving sexual feelings toward his or her parents. Adventurous ideas like these can be handled scientifically or unscientifically. Over the twentieth century, the Marxist view of history has been handled scientifically enough for it to have been disconfirmed. Too much has happened that seems to have little to do with class struggle; the ever-increasing political role of religious and cultural solidarity is an example (Huntington 1996). And capitalist societies have adapted to problems—especially economic tensions—in ways that Marxist views about politics and economics do not predict. Of course, it remains *possible* to hang onto the main principles of Marxism despite this, but fewer and fewer people handle the theory in that way anymore. Many still think that Marxism contains useful insights about economic matters, but the fundamental claims of the theory have not stood up well.

Freudianism is another matter; the ideas are still popular in some circles, but not because of success under empirical testing. Instead, the theory seems to hang around because of its striking and intriguing character, and because of a subculture in fields such as psychotherapy and literary theory which guards the main ideas and preserves them despite their empirical problems. The theory is handled very unscientifically by those groups. Freud's theory is not taken seriously by most scientifically oriented psychology departments in research universities, but it is taking a while for this fact to filter out to other disciplines.

Evolution is another big idea that can be handled scientifically or unscientifically. People (including Popper) have wondered from time to time whether evolutionary theory, or some specific version of it such as Darwinism, is testable. So they have asked, What observations would lead scientists to give up current versions of evolutionary theory? A one-line reply that biologists sometimes give to this question is "a Precambrian rabbit." An evolutionary biology textbook by Douglas Futuyma expresses the same point more soberly: finding "incontrovertibly mammalian fossils in incontrovertibly pre-cambrian rocks" would "refute or cast serious doubt on evolution" (1998, 760). The one-liner is a start, but the real situation is more complicated. So let us look at the case.

The Precambrian era ended around 540 million years ago. Suppose we found a well-preserved rabbit fossil in rocks 600 million years old. All our other evidence suggests that the only animals around then were sponges and a few other invertebrates and that mammals did not appear until over 300 million years later. Of course, a good deal of suspicion would be directed toward the finding itself. How sure are we that the rocks are that old? Might the rabbit fossil have been planted as a hoax? Remember the apparent fossil link between humans and apes that turned out to be a hoax, the Piltdown man of 1908 (see Feder 1996). Here we encounter another aspect of the problem of holism about testing—the challenging of observation reports, especially observation reports that are expressed in a way that presupposes other pieces of theoretical knowledge. This will be discussed in chapter 10. But let us suppose that all agree the fossil is clearly a Precambrian rabbit.

This finding would not be an instant falsification of all of evolutionary theory, because evolutionary theory is now a diverse package of ideas, including abstract theoretical models as well as claims about the actual history of life on earth. The theoretical models are intended to describe what various evolutionary mechanisms can do *in principle*. Claims of that kind are usually tested via mathematical analysis and computer simulation. Small-scale evolution can also be observed directly in the lab, especially in bacteria and fruit flies, and the Precambrian rabbit would not affect those results.

But a Precambrian rabbit fossil would show that *somewhere* in the package of central claims found in evolutionary biology textbooks, there are some very serious errors. These would at least include errors about the overall history of life, about the kinds of processes through which a rabbit-like organism could evolve, and about the "family tree" of species on earth. The challenge would be to work out where the errors lie, and that would require separating out and independently reassessing each of the ideas that make up the package. This reassessment could, in principle, result in the discarding of very basic evolutionary beliefs—like the idea that humans evolved from nonhumans.

Over the past twenty years or so, evolutionary theory has in fact been exposed to a huge and sustained empirical test, because of advances in molecular biology. Since the time of Darwin, biologists have been trying to work out the total family tree linking all species on earth, by comparing their similarities and differences and taking into account factors such as geographical distribution. The family tree that was arrived at prior to the rise of molecular biology can be seen summarized in various picturesque old charts and posters.

Then more recently, molecular biology made it possible to compare the DNA sequences of many species. Similarity in DNA is a good indicator of the closeness of evolutionary relationship. Claims about the evolutionary relationships between different species can be tested reasonably directly by discovering how similar their DNA is and calculating how many years of independent evolution the species have had since they last shared a "common ancestor." As this work began, it was reasonable to wonder whether the wealth of new information about DNA would be compatible or incompatible with the family tree that had been worked out previously. Suppose the DNA differences between humans and chimps had suggested that the human lineage split off from the lineage that led to chimps many hundreds of millions of years ago and that humans are very closely genetically related to squid. This would have been a disaster for evolutionary theory, one of almost the same magnitude as the Precambrian rabbit.

As it happened, the DNA data suggest that humans and chimps diverged about 4.6–5 million years ago and that chimps or pigmy chimps (bonobos) are our nearest living relatives. Prior to the DNA data, it was unclear whether humans were more closely related to chimps or to gorillas, and the date for the chimp-human divergence was much less clear. That is how the grand test of our old pre-molecular family tree has tended to go. There have been no huge surprises but lots of new facts and a lot of adjustments to the previous picture.

Further Reading

Popper's most famous work is his book *The Logic of Scientific Discovery*, published in German in 1935 and in English in 1959. The book is mostly very readable. Chapters 1–5 and 10 are the key ones. For the issues in section 4.4 above, see chapter 5 of Popper; for section 4.5, see chapter 10. A quicker and very useful introduction to Popper's ideas is the paper "Science: Conjectures and Refutations" in his collection *Conjectures and Refutations* (1963).

Newton-Smith, *The Rationality of Science* (1981), contains a clear and detailed assessment of Popper's ideas. It includes a simplified presentation of some of the technical issues surrounding corroboration that I omitted here. Salmon 1981 is an exceptionally good critical discussion of Popper's views on induction and prediction. See also Putnam 1974. Schilpp (1974) collects many critical essays on Popper, with Sir Karl's replies.

Popper's influence on biologists and his (often peculiar) ideas about evolutionary theory are discussed in Hull 1999. Horgan's book *The End of Science* (1996) contains a very entertaining interview with Popper.

5

Kuhn and Normal Science

5.1 "The Paradigm Has Shifted"

In this chapter we encounter the most famous book about science written during the twentieth century—*The Structure of Scientific Revolutions,* by Thomas Kuhn. Kuhn's book was first published in 1962, and its impact was enormous. Just about everything written about science by philosophers, historians, and sociologists since then has been influenced by it. The book has also been hotly debated by scientists themselves. But *Structure* (as the book is known) has not only influenced these academic disciplines; many of Kuhn's ideas and terms have made their way into areas like politics and business as well.

A common way of describing the importance of Kuhn's book is to say that he shattered traditional myths about science, especially empiricist myths. Kuhn showed, on this view, that actual scientific behavior has little to do with traditional philosophical theories of rationality and knowledge.

There is some truth in this interpretation, but it is often greatly exaggerated. Kuhn spent much of his time after *Structure* trying to distance himself from some of the radical views of science that came after him, even though he was revered by the radicals. And the connection between Kuhn's views and logical empiricism is actually quite complicated. For example, it comes as a surprise to many to learn that Kuhn's book was published in a series organized and edited *by* the logical empiricists; *Structure* was published as part of their International Encyclopedia of Unified Science series. As a matter of historical fact, though, there is no denying that this was something of a "Trojan horse" situation. Logical empiricism was widely perceived as being seriously damaged by Kuhn.

I said above that some of Kuhn's ideas and terms have made their way into areas far from the philosophy of science. The best example is Kuhn's use of the term "paradigm." Here is a passage from Tom Wolfe's 1998 novel,

A Man in Full. Charlie Croker, a real estate developer who has debt problems, is talking with his financial adviser, Wismer ("Wiz") Stroock.

> "I'm afraid that's a sunk cost, Charlie," said Wismer Stroock. "At this point the whole paradigm has shifted."
>
> Charlie started to remonstrate. Most of the Wiz's lingo he could put up with, even a "sunk cost." But this word "paradigm" absolutely drove him up the wall, so much so that he had complained to the Wiz about it. The damned word meant nothing at all, near as he could make out, and yet it was always "shifting," whatever it was. In fact, that was the only thing the "paradigm" ever seemed to do. It only shifted. But he didn't have enough energy for another discussion with Wismer Stroock about technogeekspeak. So all he said was:
>
> "OK, the paradigm has shifted. Which means what?" (71)

This sort of talk derives completely from Kuhn. But what *is* a paradigm? The short answer is that a paradigm, in Kuhn's theory, is a whole way of doing science, in some particular field. It is a package of claims about the world, methods for gathering and analyzing data, and habits of scientific thought and action. In Kuhn's theory of science, the big changes in how scientists see the world—the "revolutions" that science undergoes every now and then—occur when one paradigm replaces another. Kuhn argued that observational data and logic alone cannot force scientists to move from one paradigm to another, because different paradigms often include within them different rules for treating data and assessing theories. Some people have interpreted Kuhn as claiming that changes between paradigms are completely irrational, but Kuhn definitely did not believe that. Instead, Kuhn had a complicated and subtle view about the roles of observation and logic in scientific change.

In a passage like the Tom Wolfe one above, "paradigm" is used in a looser way derived from its role in Kuhn's theory of science. A paradigm in this sense means something like a way of seeing the world and interacting with it.

Kuhn did not invent the word "paradigm." It was an established term, which meant (roughly) an illustrative example of something, on which other cases can be modeled. Kuhn discusses this original meaning in *Structure* (1996, 23). And although Kuhn's theory is the inspiration for all the talk about paradigm shifts that one hears, Kuhn only occasionally used the phrase "paradigm shift." More often he talked about paradigms changing or being replaced. Whichever term one uses, though, Kuhn's theory was itself something like a paradigm change in the history and philosophy of science. Nothing has been the same since.

5.2 Paradigms: A Closer Look

A moment ago I said that a paradigm in Kuhn's theory is a package of claims about the world, methods for gathering and analyzing data, and habits of scientific thought and action. However, it is more accurate to say that this is *one* sense in which Kuhn used the term "paradigm." In *Structure,* the term is used in several different ways; one critic counted as many as twenty-one different senses (Masterman 1970). Kuhn later agreed that he had used the word ambiguously, and throughout his career he kept fine-tuning this and other key concepts. To keep things simple, though, in this book I will recognize two different senses of the term "paradigm."

The first sense, which I will call the *broad* sense, is the one I described above. Here, a paradigm is a package of ideas and methods, which, when combined, make up both a view of the world and a way of doing science. When I say "paradigm" in this book without adding "broad" or "narrow," I mean this broad sense. But there is also a narrower sense. According to Kuhn, one key *part* of a paradigm in the broad sense is a specific *achievement,* or an *exemplar.* This achievement might be a strikingly successful experiment, such as Mendel's experiments with peas, which eventually became the basis of modern genetics. It might be the formulation of a set of equations or laws, such as Newton's laws of motion or Maxwell's equations describing electromagnetism. Whatever it is, this achievement is a source of inspiration to others; it suggests a way to investigate the world. Kuhn often used the term "paradigm" just for a specific achievement of this kind. I will call these achievements paradigms in the *narrow* sense. So paradigms in the broad sense (whole ways of doing science) include within them paradigms in the narrow sense (examples that serve as models, inspiring and directing further work). Kuhn himself did not use this "narrow/broad" terminology, but it is helpful. When Kuhn first introduced the term "paradigm" in *Structure,* he defined it in the narrower sense. But in much of his writing, and in most of the work written after *Structure* using the term, the broad sense is intended.

Kuhn used the phrase "normal science" for scientific work that occurs within the framework provided by a paradigm. A key feature of normal science is that it is *well organized.* Scientists doing normal science tend to agree on which problems are important, on how to approach these problems, and on how to assess possible solutions. They also agree on what the world is like, at least in broad outlines. A scientific revolution occurs when one paradigm breaks down and is replaced by another.

This initial sketch is enough for us to go straight to some central points about the message of Kuhn's book.

The first point can be approached via a contrast with Popper. For Popper, science is characterized by *permanent openness,* a permanent and all-encompassing *critical stance,* even with respect to the fundamental ideas in a field. Other empiricist views will differ on the details here, but the idea of science as featuring permanent openness to criticism and testing is common to many versions of empiricism. Kuhn disagreed. He argued that it is *false* that science exhibits a permanent openness to the testing of fundamental ideas. Not only that, but science would be *worse off* if it had the kind of openness that philosophers have treasured.

The second point concerns scientific change. Here again a contrast with Popper is convenient. For Popper, all science proceeds via a single process, the process of conjecture and refutation. There can still be episodes called "revolutions" in such a view, but revolutions are just different in degree from what goes on the rest of the time; they involve bigger conjectures and more dramatic refutations. For Kuhn, there are two distinct kinds of scientific change: change within normal science, and revolutionary science. (These are bridged by "crisis science," a period of unstable stasis.) These two kinds of change have very different epistemological features; when we try to apply concepts such as *justification, rationality,* and *progress* to science, according to Kuhn we find that normal and revolutionary science have to be described very differently. Within normal science, there are clear and agreed-upon standards for the justification of arguments; within revolutionary science there are not. Within normal science there is clear progress; within revolutionary science it is very hard to tell (and it is hard to even interpret the question). Because revolutions are essential to science, the task of describing rationality and progress in science *as a whole* becomes very complicated.

So Kuhn is first making some claims about how science actually operates, and then drawing philosophical conclusions from those claims. Even if we leave aside the details of Kuhn's claims, this strategy of argument was controversial and influential. Kuhn addressed philosophical questions about reason and evidence via an examination of history. As we saw in chapter 2, the logical empiricists made a sharp distinction between questions about the history and psychology of science, on the one hand, and questions about evidence and justification, on the other. Kuhn was deliberately mixing together things that the logical empiricists had insisted should be kept apart. One of the reasons that Kuhn was interpreted as a "destroyer" of logical empiricism was that Kuhn's work seemed to show how interesting it is to connect philosophical questions about science with questions about the history of science. Kuhn seemed to open up an exciting new way of approaching a set of problems that the logical empiricists were approaching in a very abstract manner. And although I emphasized in earlier chapters some appealing parts of the logical empiricist approach, I agree with Kuhn about the useful role of history in addressing philosophical questions about science.

Before we go deeper into the details of Kuhn's view, there is one other preliminary point to make. This has to do with a question that one should always ask when thinking about Kuhn's theory and other theories like it. The question is, Which parts of the theory are just *descriptive,* and which are *normative*? That is, when is Kuhn just making a claim about how things *are,* and when is he making a value judgment, saying how they *should* be? Kuhn certainly accepted that he was making some normative claims (1996, 8). Some commentators were critical of Kuhn, however, because it's often hard to tell when he is just "saying how things are" and when he is making claims about good and bad science. My own interpretation of Kuhn stresses the normative element in his work. I think Kuhn had a very definite picture of how science should work and of what can cause harm to science. In fact, it is here that we find what I regard as the most fascinating feature of *The Structure of Scientific Revolutions.* This is the relationship between

1. Kuhn's constant emphasis on the arbitrary, personal nature of factors often influencing scientific decisions, the rigidity of scientific indoctrination of students, the "conceptual boxes" that nature gets forced into by scientists . . . , and
2. Kuhn's suggestion that these features are actually the *key to science's success*—without them, there is no way for scientific research to proceed as effectively as it does.

Kuhn is saying that without the factors referred to in (1), we would not have the most valuable and impressive features of science. But how can this be? How can features that look like failings and flaws actually *help* science? How can it *help* science for decisions to be made on the basis of anything other than what the data say? To answer these questions, we need to look more closely at the details of Kuhn's story about scientific change.

5.3 Normal Science

Normal science is work inspired by a striking achievement that provides a basis for further work (a paradigm in the narrow sense). Kuhn does not think that all science needs a paradigm. Each scientific field starts out in a state of "pre-paradigm science." During this pre-paradigm state, scientific work can go on, but it is not well organized and usually not very effective.

At some point, however, some striking piece of work appears. This achievement is taken to provide insight into the workings of some part of the world, and it supplies a model for further investigation. This achievement is so impressive that a tradition of further work starts to grow up around it. The field has its first paradigm.

What are some examples of paradigms? Kuhn gave examples from physics and chemistry, such as Newton's and Einstein's paradigms. Here I will mention two cases from other fields. Within psychology around the middle of the twentieth century, a great deal of work was based upon the behaviorist approach of B. F. Skinner. Two basic principles of Skinnerian behaviorism are (1) that learning is basically the same in humans, rats, pigeons, and other animals and (2) that learning proceeds by reinforcement—behaviors followed by good consequences tend to be repeated, while behaviors followed by bad consequences tend not to be repeated. Along with these principles, the Skinnerian paradigm included a set of experimental tools, such as standardized boxes in which pigeons made choices in response to stimuli by pecking lighted keys. It also included statistical techniques used to analyze data and various habits and skills for working out relevant and interesting experiments.

Here is an example from biology. Modern molecular genetics is based on a set of principles such as the following: (1) genes are made of DNA (in all organisms except some viruses, which have RNA genes), (2) genes have their effects by producing protein molecules and regulating other genes, and (3) nucleic acids (DNA and RNA) specify the structure of proteins but not vice versa. This last principle is often called "the central dogma." Along with these theoretical claims, molecular genetics includes a set of techniques for sequencing genes, for producing and studying mutations, for analyzing the similarity of different genes, and so on.

For Kuhn, a scientific field usually has only one paradigm guiding it at any particular time. Sometimes Kuhn wrote as if this were true by definition—a *field* being defined as an area of scientific investigation unified by a single paradigm. This led him to divide some scientific fields up more finely than is usual. Kuhn does allow that occasionally a field can be governed by several related paradigms, but this is rare. In general, a key part of Kuhn's theory is the principle *one paradigm per field per time*.

A paradigm's role is to *organize* scientific work; the paradigm coordinates the work of individuals into an efficient collective enterprise. A key feature that distinguishes normal science from other kinds of science for Kuhn is the absence of debate about fundamentals. Because scientists doing normal science agree on these fundamentals, they do not waste their time arguing about the most basic issues that arise in their field. Once biologists agree that genes are made of DNA, they can focus and coordinate their work on how specific genes affect the characteristics of plants and animals. Once chemists agree that understanding chemical bonding is understanding the interactions between the outer layers of electrons within different atoms, they can work together to investigate when and how particular reactions will occur. Kuhn places great emphasis on this "consensus-forging" role of paradigms. He argues that without it, there is no chance for scientists to achieve a really detailed and deep understanding of phenomena. Detailed work and revealing discoveries require cooperation and consensus. Cooperation and consensus require *closing off debate about fundamentals.*

As usual, we should be careful to distinguish between the descriptive and the normative here. Kuhn certainly claims that normal science does close off debate about fundamentals. But does he go beyond that and claim this is something that normal science *should* do? I think it is clear that he does (see Kuhn 1996, 24–25, 65), but these issues are controversial.

If Kuhn does make a normative claim here, then we see an important contrast with Popper. Although Popper can certainly allow that not everything can be criticized at once within science, Popper's view does hold that a good scientist is permanently open-minded with respect to *all* issues in the field in which he or she is working, even the very basic issues. Any "closing off" of debate is bad news according to Popper. Popper criticized Kuhn explicitly on this point; Popper said that although "normal science" of Kuhn's kind does occur, it is a bad thing that it does (1970).

What is the work of a good normal scientist like? Kuhn describes much of the work done in normal science as "puzzle-solving." The normal scientist tries to use the tools and concepts provided by the paradigm to describe, model, or create new phenomena. The "puzzle" is trying to get a new case to fit smoothly into the framework provided by the paradigm. Kuhn used the term "puzzle" rather than "problem" for a reason. A puzzle is something we have not yet solved but which we think does have a solution. A problem might, for all we know, have no solution. Normal science tries to apply the concepts provided by a paradigm to issues that the paradigm suggests *should* be soluble. Part of the guidance provided by a paradigm is guiding the selection of good puzzles.

The term "puzzle" also seems to suggest that the work is in some way insignificant or trivial. Here again, Kuhn intends to convey a precise message with the term. A normal scientist *does,* Kuhn thinks, spend a lot of time on topics that look insignificant from the outside. (He even uses the term "minuscule" [1996, 24].) But it is this close attention to detail—which only the well-organized machine of normal science makes possible—that is able to reveal *deep* new facts about the world. I think Kuhn felt a kind of awe at the ability of normal science to home in on topics and phenomena that look insignificant from outside but which turn out eventually to have huge importance. And although the normal scientist is not *trying* to find phenomena that lead to paradigm change—far from it!—these detailed discoveries often contain the seeds of large-scale change and the destruction of the paradigm that produced them.

5.4 Anomaly and Crisis

I said that a central feature of normal science, for Kuhn, is that the fundamental ideas associated with a paradigm are not debated. Fundamental principles are insulated from refutation. Normal scientists spend their time trying to extend the paradigm, theoretically and experimentally, to deal with new cases. When there is a failure to get the results expected, the good normal scientist reacts by trying to work out what mistake she or he has made. The proverb "only a poor workman blames his tools" applies. The normal scientist should take failure as a challenge.

Kuhn accepts that theories are sometimes refuted by observation; within normal science, hypotheses are refuted (and confirmed) all the time. The paradigm supplies principles for making these decisions. But throwing out an entire paradigm is much more difficult. According to Kuhn, the rejection of a paradigm happens only when (1) a critical mass of anomalies has arisen and (2) a rival paradigm has appeared. For now we will look just at the first of these—the accumulation of a critical mass of anomalies.

An "anomaly" for Kuhn is a puzzle that has resisted solution. Kuhn holds that all paradigms face some anomalies at any given time. As long as there are not too many of them, normal science proceeds as usual and scientists regard them as a challenge. But the anomalies tend to accumulate. Sometimes a single one becomes particularly prominent, by resisting the efforts of the best workers in the field. Eventually, according to Kuhn, the scientists start to lose faith in their paradigm. The result is a *crisis*.

Crisis science, for Kuhn, is a special period when an existing paradigm has lost the ability to inspire and guide scientists, but when no new paradigm has emerged to get the field back on track. The transition to a crisis is almost like a phase transition, like the change of a substance from solid to liquid during melting. For whatever reason, the scientists in a field lose their confidence in the paradigm. As a consequence, the most fundamental issues are back on the table for debate. Amusingly, Kuhn even suggests that during crises scientists tend to suddenly become interested in philosophy, a field that he sees as quite useless for normal science.

I used the term "critical mass" of anomalies to describe the trigger for a crisis. This atomic-age metaphor is appropriate in several ways. In particular, I use it here to suggest that Kuhn sees the breakdown of a paradigm as something that is part of the "proper functioning" of science, though it does not feel that way to the scientists involved. Normal science is structured in a way that makes its own destruction inevitable, but only in response to the right stimulus. The "right stimulus" is the appearance of problems that are deep rather than superficial, problems that reveal a real inadequacy in the paradigm. Because normal scientists will tolerate a good deal of temporary trouble without abandoning normal science—they will blame the failure on themselves, at least for a while—a paradigm does not break down easily. But when the right stimulus comes, the paradigm will break down. In this way, a paradigm is like a well-shielded and well-designed bomb. A bomb is supposed to blow up; that is its function. But a bomb is not supposed to blow up at any old time; it's supposed to blow up in very specific circumstances. A well-designed bomb will be shielded from minor buffets. Only a very specific stimulus will trigger the explosion.

Some might find this militaristic analogy unpleasant, but I think it captures a lot of what Kuhn says. Kuhn's story is guided by his claim that all paradigms constantly encounter anomalies. For a Popperian view, or for other simpler forms of empiricism, these anomalies should count as "refutations" of the theory. But Kuhn thinks that science *does* not treat these constantly arising anomalies as refutations, and also that it *should* not. If scientists dropped their paradigms every time a problem arose, they would never get anything done.

Much of the secret of science, for Kuhn, is the remarkable balance it manages to strike between being *too* resistant to change in basic ideas, and not being resistant enough. If the simplest form of empiricist thinking prevailed, people would throw ideas away too quickly when unexpected observations appeared, and chaos would result. Ideas need some protection, or they can never be properly developed. But if science was completely unresponsive to empirical failures, conceptual advance would grind to a halt. For Kuhn, science seems to get the balance just right. And this delicate balance is not something we can describe in terms of a set of explicit rules. It exists implicitly in the social structures and transmitted traditions of scientific behavior, and in the quirks of the scientific mind.

These ideas about the balance that makes science work are an important challenge to empiricism, at least in its simpler forms. The idea that a willingness to revise ideas in response to observation *can go too far* is unexpected from the point of view of empiricist philosophy. And Kuhn supported this claim with a mass of evidence from the history of science.

So far we have gone from pre-paradigm science, through normal science, to crisis. The next stage in Kuhn's story is revolution. But before we get there, I will make some summary remarks about Kuhn's theory of normal science.

5.5 Wrap-up of Normal Science

Let's sum up what we have so far. Paradigms function to organize scientific work. Normal science is work aimed at extending and refining the paradigm. A good normal scientist is committed to the paradigm and does not question it. Normal scientists extend their paradigm both theoretically and experimentally. Anomalies inevitably arise, however, and eventually these reach a kind of critical mass, at which point scientists lose faith in the paradigm and the field plunges into a state of crisis.

We have not yet reached the most controversial part of Kuhn's theory, but are there any problems with what we have so far? One problem comes from Kuhn's insistence that, except in unusual cases, a scientific field has one paradigm per field per time. Kuhn held that, in general, a single paradigm will dominate its field. He did not think that two or three separate and competing paradigms could normally coexist. Many critics have thought Kuhn was wrong about this, both in the cases of physics and chemistry, which he discussed extensively, and, even more so, for areas like biology and psychology, which he did not often discuss. We will come back to this issue in chapter 7.

Secondly, Kuhn exaggerates the degree of commitment that a normal scientist does and should have to a paradigm. Kuhn describes the attitude of a normal scientist in very strong terms. Scientific education is a kind of "indoctrination," which results in scientists having a deep "faith" in their paradigm. As a description of how science actually works, this seems exaggerated. Sometimes there is a faithlike commitment, but sometimes there is not. Many scientists are able to say that they always *work* within a paradigm, for practical reasons, while being very aware of the possibility of error and the eventual replacement of their framework. One of the ironies of Kuhn's influence is that his book might have weakened the faith of some normal scientists, even though Kuhn thought that normal scientists should have a deep faith in their paradigms!

Leaving aside the factual issue of whether a tenacious commitment to a paradigm is what we generally find, we should also ask about Kuhn's belief that this strong commitment is a good thing. For Kuhn, the great virtue of normal science is its organized, coordinated structure, a structure that results in precision and efficiency. Unless debate about fundamentals is closed off, this precision and efficiency will be reduced. A key contrast here is with

Popper, who insists on permanent open-mindedness. For Kuhn, a constant questioning and criticism of basic beliefs is liable to result in chaos—in the partially "random" fact-gathering and speculation that we see in pre-paradigm science. But here again, Kuhn probably goes too far. He does not take seriously the possibility that scientists could *agree* to work together in a coordinated way, not wasting time on constant discussion of fundamental issues, while retaining a *cautious* attitude toward their paradigm. Surely this is possible.

To close this chapter, I will describe an example that is very far from Kuhn's usual cases—an example that is on the periphery of science, but which I think illustrates Kuhn's insights well. During the 1980s and 1990s, there was a lot of excitement about a new field known as "Artificial Life." The aim was to use computers to model the most basic features of living systems, in such a way that it might eventually be reasonable to say that the artificial systems *were* alive. I went to several "Alife" conferences during this time and watched the field develop. At the time that I write this, the field seems to have ground to a halt. Maybe it will revitalize. But the failure of the field in recent years seems to me to involve some very "Kuhnian" reasons.

During the heyday of the movement, two or three pieces of work appeared that were strikingly successful. Perhaps the most impressive of all was Tom Ray's Tierra project, in which Ray was able to create open-ended evolution among self-replicating programs in a computer (Ray 1992). Work by Chris Langton and Steven Wolfram on the mathematical analysis of "cellular automata," simple systems in which local interactions between elements give rise to global, self-sustaining patterns, might be another case. And closer to the borderline with mainstream biology, there was Stuart Kaufmann's work on "the origins of order" in complex systems (Kauffman 1993).

All of this was impressive work, and it pointed the way forward to a consolidation of what these imaginative individuals had done. But the consolidation never happened. At each conference I went to, the larger group of people involved all seemed to want to do things from scratch, in their own way. Each had his or her own way of setting up the issues. There was not nearly enough work that *built on* the promising beginnings of Ray and others. The field never made a transition into anything resembling normal science. And it has now ground to a halt.

Another reason for the breakdown also relates to Kuhn. The field of Alife suffered from a kind of "premature commercialization." It was realized early on that some of the work had great potential for animation and other kinds of commercial art. At some Alife events, the climax of each talk seemed to be not some new theoretical idea, but a dramatic video. (I even heard speakers half-apologize before the video, as if they knew this was somehow not the right emphasis for their work.) For Kuhn, science depends on the good normal scientist's keen interest in puzzle-solving for its own sake. Looking outside the paradigm too often to applications and external rewards is not good for normal science.

Further Reading

Lakatos and Musgrave's collection *Criticism and the Growth of Knowledge* (1970) contains an excellent set of essays on Kuhn. A more recent edited collection is Horwich, *World Changes* (1993).

Kuhn's collection of essays *The Essential Tension* (1977b) is an important extra source. Kuhn also wrote two historical books (1957, 1978). His later essays have been collected in *The Road since Structure* (2000).

Levy 1992 is a readable survey of Alife work. Many of the best Alife papers may be found in the collection *Artificial Life II* (Langton et al. 1992).

6

Kuhn and Revolutions

I have argued so far only that paradigms are constitutive of science. Now I wish to display a sense in which they are constitutive of nature as well.

THOMAS KUHN, *Structure*

"Look," Thomas Kuhn said. The word was weighted with weariness, as if Kuhn was resigned to the fact that I would misinterpret him, but he was still going to try—no doubt in vain—to make his point. "Look," he said again. He leaned his gangly frame and long face forward, and his big lower lip, which ordinarily curled up amiably at the corners, sagged. "For Christ's sake, if I had my choice of having written the book or not having written it, I would choose to have written it. But there have certainly been aspects involving considerable upset about the response to it."

JOHN HORGAN, *The End of Science*

6.1 Considerable Upset

The most famous, most striking, and most controversial parts of Kuhn's book were his discussions of scientific revolutions. They are the topic of this chapter. Why have two chapters on Kuhn? One reason is the continuing importance and great subtlety of his book. Another is that while the discussions of revolution are the most famous parts of the book, Kuhn's analysis of normal science is just as important—and perhaps of more enduring significance. Sometimes it gets lost in the excitement about revolutions: hence chapter 5.

Kuhn argued that some periods of scientific change involve a fundamentally different kind of process from what we find in normal science. The revolutionary periods see a breakdown of order and a questioning of the rules of the game, and they are followed by a process of rebuilding that can create fundamentally new kinds of conceptual structures. Revolutions involve a breakdown, but they are *essential* to science as we know it. They have a "function," Kuhn often said, within the totality of science. The special features we associate with science arise from the combination and interaction of two different kinds of activity—the orderly, organized, disciplined process of normal science, and the periodic breakdowns of order found in revolutions. These two processes happen in sequence, within each scientific field. Science as a whole is a result of their interaction and of nothing less.

Kuhn seemed to divide science into units with strange boundaries between them. Looking *within* a period of normal science, you can easily distinguish good work from bad, rational moves from irrational, big problems from small problems, and so on. Progress is evident as time goes by. But all this ends with a revolution. In a scientific revolution, as in a political one, rules break down and have to be rebuilt afresh. If you look at two pieces of scientific work *across* a revolutionary divide, it will not be clear whether there has been progress from earlier to later. It might not even be clear how to compare the theories or pieces of work *at all*—they may look like fundamentally different kinds of intellectual activity. The people on different sides of the divide will be "speaking different languages." In the climax of his book, Kuhn says that workers in different paradigms are living in different worlds.

6.2 Revolutions and Their Aftermath

A revolution is a kind of discontinuity in the history of a scientific field. The suggestion that science has two modes of change, one of them dramatic and abrupt, does not itself have big consequences for philosophy, though it is interesting. The big issues depend upon what the two modes of change are like. And here there are two sets of issues that caused intense discussion. The first is *how* revolutions occur—what goes on within them. The second has to do with the relations between what we have before and what we have after a revolution.

How do revolutions occur? We finished the previous chapter describing the transition from normal science to crisis. In Kuhn's story, large-scale scientific change usually requires both a crisis *and* the appearance of a new candidate paradigm. A crisis alone will not induce scientists to regard a large-scale theory or paradigm as "falsified." We do not find pure falsifications, rejections of one paradigm without simultaneous acceptance of a new one. Rather, the rejection of one paradigm accompanies the acceptance of another. But also, the switch to a new paradigm does not occur just because a new idea appears which looks better than the old one. Without a crisis, scientists will not have any motivation to consider radical change.

All of Kuhn's claims about what follows what in scientific change tend to be qualified; he is describing the central and characteristic patterns of change, not every case without exception. But the idea that revolutions generally require crises raised some hard historical issues. Was there a crisis in the state of astronomy before Copernicus, or in biology before Darwin? Was there a state of disorder following an earlier period of confident work? Maybe. But taking another biological example, if the appearance of genetics as a science around 1900 was a revolution, it is *very* hard to find a crisis in the work on inheritance that preceded it. (Maybe Kuhn would regard this as a transition from pre-paradigm science to normal science, though that could not be said about most of biology around that time.) Some other twentieth-century revolutions, such as the molecular revolution in biology, seem even less crisis-induced.

In his 1970 "Postscript" to *Structure,* Kuhn qualified his claims about the role of crisis (181). He still maintained that crises are the "usual prelude" to revolutions. But even that claim is controversial. Kuhn's emphasis on crises sometimes seems driven more by the demands of his hypothesized mechanism for scientific change than by the historical data; Kuhn's story demands crises because only a crisis can loosen the grip of a paradigm and make people receptive to alternatives.

Suppose we do have a crisis, a period full of confusion and strange guests in the philosophy department. Then a new candidate paradigm appears, precipitating a revolution. Using my distinction from the previous chapter, what initially appears is a new paradigm in the *narrow* sense, an achievement that begins to inspire people and seems to point the way forward. More specifically, what is usually involved is that the new work appears to solve one or more of the problems that prompted the crisis in the old paradigm. The sudden appearance of problem-solving power is the spark to the revolution. Kuhn did not think these processes could be described by an explicit philosophical theory of evidence and testing. Instead, we should think of the shift to a new paradigm as like a "conversion" phenomenon, or like a gestalt switch. Kuhn also argued that revolutions are capricious, disorderly events. They are affected by idiosyncratic personal factors and accidents of history.

One reason for the disorderly character of revolutions is that some of the principles by which scientific evidence is assessed are themselves liable to be destabilized by a crisis, and they can change with a revolution. Kuhn did not argue that traditional philosophical ideas about how theories should relate to evidence are completely misguided. He made it clear in his later work that there are some core ways of assessing theories that are common to all paradigms (1977c, 321–22). Theories should be predictively accurate, consistent with well-established theories in neighboring fields, able to unify disparate phenomena, and fruitful of new ideas and discoveries. These principles, along with other similar ones, "provide *the* shared basis for theory choice" (322). (I should note that some commentators think these later essays change, rather than clarify, the views presented in *Structure*.)

But Kuhn thought that when these principles were expressed in a broad enough way to be common across all of science, they would be so vague that they would be powerless to settle hard cases. Also, these goals must often be traded off against each other; emphasizing one will require downplaying another. *Within* a single paradigm, more precise ways of assessing hypotheses will operate. These will include sharper versions of the common principles listed above, but these sharper versions will not really be explicit "principles" anymore. Instead they will be more like habits and values, aspects of the shared mind-set of normal scientists imparted to them by their common training and common activities. There will also be some variation *within* normal science in how these principles are understood and acquired—Kuhn came to see this diversity as a strength of scientific communities as well. But the most important point here is that these sharper, more definite ways of assessing ideas are liable to change in the course of a revolution. In the next section I will give an example of this phenomenon.

So we have two kinds of scientific change in Kuhn's picture, neither of which is what empiricist philosophies of science might have led us to expect. Change within normal science is orderly and responsive to evidence—but normal science works via a closing of debate about fundamental ideas. The other kind of change—revolutionary change—does involve challenges to fundamentals, but these are episodes in which the orderly assessment of ideas breaks down. Displays of problem-solving power have a key role in these fundamental transitions between paradigms, but the shifts also involve sudden gestalt switches and leaps of faith.

In Kuhn's treatment of revolutionary change, the distinction between descriptive and normative issues is very important. Kuhn uses language that suggests that not only are revolutions bound to happen, but they have a positive role in science. They are part of what makes science so powerful as a means for exploring the world (a "supremely efficient instrument" [1996, 169]). Different interpreters have very different reactions to this kind of talk. Some regard it as colorful and not essential to Kuhn's general message. I have the opposite view; I think this is central to Kuhn's overall picture. Science for Kuhn is a social mechanism that combines two capacities. One is the capacity for sustained, cooperative work. The other is science's capacity to partially break down and reconstitute itself from time to time. When a paradigm runs out of steam, there is nothing within the community that could reliably give science a set of directions for *orderly* movement toward a new paradigm. Instead, the goals of science are best served at these special times by a *disorderly* process, in which even very basic ideas are put back on the table for discussion, and a new direction eventually emerges from the chaos. This sounds strange, but I think it was Kuhn's picture.

6.3 Incommensurability, Relativism, and Progress

Kuhn said that revolutions have a "non-cumulative" nature; this is essential to his claims about the large-scale historical patterns in science. There is no steady buildup of some useful commodity like truth as science goes along. Instead, according to Kuhn, in a revolution you always *gain* some things and *lose* some things. Questions that the old paradigm answered now become puzzling again, or they cease to be questions. So we might want to ask, Do we usually gain more than we lose? In at least the middle chapters of his book, Kuhn seems to think there is no way to answer this question in an unbiased way (1996, 109, 110). Of course it will *feel* like we have gained more than we've lost, or we would not have had the revolution at all. But that does not mean that there is some unbiased way of comparing what we had before with what we have after.

This question connects us to one of the most famous topics in Kuhn's work, the idea that different paradigms in a field are *incommensurable* with each other.

What does "incommensurable" mean here? Most literally, it means not comparable by use of a common standard or measure. This idea needs to be carefully expressed, however. Two rival paradigms can be compared well enough for it to be clear that they are incompatible, that they are *rivals*. And people working within any one paradigm will have no problem saying why their paradigm is superior to the other, by citing key differences in what can be explained and what cannot. But these comparisons will be compelling only to those inside the paradigm from which the claim of superiority is being made. If we look down "from above" on two people who work within different paradigms who are arguing about which is better, it will often appear that the two people are talking past each other.

There are two reasons for this—there are (roughly speaking) two aspects of the problem of incommensurability. First, people in different paradigms will not be able to fully *communicate* with each other; they will use key terms in different ways and in a sense will be speaking slightly different languages. Second, even when communication is possible, people in different paradigms will use different *standards of evidence and argument.* They will not agree on what a good theory is supposed to do.

First let us look at the issues involving language. Here Kuhn's claims depend on a holistic view about the meaning of scientific language. Each term in a theory derives its meaning from its place in the whole theoretical structure. Two people from different paradigms might seem to use the same word—"mass" or "species"—but the meanings of these terms will be slightly different because of their different roles in the two rival theories.

Here I said "*slightly* different." Kuhn insisted he had a moderate view. Some critics have argued that a holistic view of meaning really has no way to make sense of these differences of degree, because it is not possible to say whether two terms have a "similar" role within two very different theoretical networks (Fodor and LePore 1992). So the critics argue that when holists about scientific language talk about "partial" communication and "slight" differences in the meanings of words, they are bluffing in order to hide the impossibly radical nature of their views.

Neither the holists nor anyone else has had much success in developing a good theory of meaning for scientific language. This is a confusing and unresolved area. However, a different kind of criticism of Kuhn is possible here. If incommensurability of meanings is real, as Kuhn says, then it should be visible in the history of science. So those who study the history of science should be able to find many examples of the usual *signs* of failed communication—confusion, correction, a sense of failure to make contact. Although I am not a historian of science, my impression is that historians have not found many examples of failed communication in crucial debates across rival paradigms. Scientists are often adept at "scientific bilingualism," switching from one framework to another. And they are often able to improvise ways of bridging linguistic gaps, much as traders from different cultures are able to, by improvising "pidgin" languages (Galison 1997). Scientists often willfully misrepresent each other's claims, in the service of rhetorical points, but that is not a case of failed comprehension or communication.

The other form of incommensurability is much more important. This is incommensurability of *standards.* Here Kuhn argued that paradigms tend to bring with them their own standards for what counts as a good argument or good evidence.

This topic was introduced in the previous section. There I said that Kuhn thought that although all scientific work is responsive to some broad principles of theory choice, the detailed standards for assessing ideas will often be internal to paradigms and liable to change with revolutions. For Kuhn, "paradigms provide scientists not only with a map but also with some of the directions essential for map-making" (1996, 109).

One of Kuhn's most interesting examples of this phenomenon involves the role of *causal explanation*. Should a scientific theory be required to make causal sense of why things happen? Should we always hope to understand the *mechanisms* underlying events? Or can a theory be entirely acceptable if it gives a mathematical formalism that describes phenomena without making causal sense of them? A famous example of this problem concerns Newton's theory of gravity. Newton gave a mathematical description of gravity—his famous inverse square law—but did not give a mechanism for how gravitational attraction works. Indeed, Newton's view that gravity acts instantaneously and at a distance seemed to be extremely hard to supplement with a mechanistic explanation. Was this a problem with Newton's theory, or should we drop the demand for a causal mechanism and be content with the mathematical formalism? Would it be scientifically acceptable to regard gravity as just an "innate" power of matter that follows a mathematical law? People argued about this a good deal in the early eighteenth century. Kuhn's view is that there is no general answer to the question of whether scientific theories should give causal mechanisms for phenomena; this is the kind of principle that will be present in one paradigm and absent from another.

During the earliest part of the twentieth century, there was a similar, although smaller-scale, debate within English biology. In the latter part of the nineteenth century, a group of biologists called the "Biometricians" had formulated a mathematical law that they thought described inheritance. They had no mechanism for how inheritance works, and their law did not lend itself to supplementation with such a mechanism. In 1900 the pioneering work done by Mendel in the mid-nineteenth century was rediscovered, and the science of genetics was launched. For about six years, though, the Biometricians and the Mendelians conducted an intense debate about which approach to understanding inheritance was superior. One of the issues at stake was what *kind* of theory of inheritance should be the goal. The Biometricians thought that a mathematically formulated law was the right goal, while William Bateson, in the Mendelian camp, argued that understanding the mechanism of inheritance was the goal. In the short term, the Mendelians won the battle. In time the two approaches were married; modern biology now has both the math and the mechanism. But during the battle there was considerable argument about what a good scientific theory should do (Provine 1971; MacKenzie 1981).

So although Kuhn's claims about linguistic incommensurability were overstated, I agree with him that incommensurability of standards is a real and interesting problem.

Kuhn's discussion of incommensurability is the main reason why his view of science is often referred to as "relativist." Kuhn's book is often considered one of the first major steps in a tradition of work in the second half of the twentieth century that embraced relativism about science and knowledge. Kuhn himself was shocked to be interpreted this way.

But what *is* relativism? This is a chaotic area of discussion. Roughly speaking, relativist views tend to hold that the truth or justification of a claim, or the applicability of a rule or standard, depends on one's situation or point of view. Such a claim might be made generally ("all truth is relative") or in a more restricted way, about art, morality, good manners, or some other particular domain. The "point of view" might be that of an individual, a society, or some other group.

If people *differ* about the facts or the proper standards in some domain, that itself does not imply that relativism is true for that domain; some of the people might just be wrong. It is also important that if someone holds that moral rightness or good reasoning "depends on context," that need not be a form of relativism, although it might be. This is because a single set of moral rules (or rules of reasoning) might have *built into them* some sensitivity to circumstances. A set of moral rules might say, "If you are in circumstances X, you should do Y." That is not relativism, even though not everyone might be in circumstances X.

In this discussion we are mostly concerned with relativism applied to *standards*. More specifically, we are concerned with standards governing reasoning, evidence, and the justification of beliefs. And the "point of view" here is that of the users of a paradigm.

Is Kuhn a relativist with regard to these matters? The answer is that it's complicated. Kuhn had a subtle view that is hard to categorize. There is no simple answer, and I doubt that everything Kuhn said on the topic can be fitted together consistently. The issue of relativism in Kuhn is also bound up with the question of how to understand scientific *progress*, something Kuhn struggled with in the final pages of his book.

As we have seen, Kuhn argued that different paradigms often carry with them different rules for assessing theories and different standards for good and bad scientific work. So far, this does not tell us whether Kuhn was a relativist about these standards. But Kuhn also argued that the paradigms we have in science now are *not* closer than earlier paradigms to an "ideal" or "perfect" paradigm. Science is not heading toward a final paradigm that is superior to all others. We cannot say that there is, in principle, an ideal paradigm that contains methodological principles that are entitled to govern all of science, even though we do not have this paradigm yet.

This seems to be taking us close to a relativist view about the standards that are not shared across paradigms. But Kuhn said some rather different things in the final, somewhat puzzling, pages of *Structure.* There he said that our present paradigms have *more problem-solving power* than earlier paradigms did. This claim was made when Kuhn confronted the question of how to understand *progress* in science.

Kuhn gave two very different kinds of explanation for the apparent large-scale progress we see in science, and these two explanations are intertwined with each other in complex ways. Kuhn's first form of explanation was a kind of "eye of the beholder" explanation. Science will inevitably *appear* to exhibit progress because each field has one paradigm per time, the victors after each revolution will naturally view their victory as progressive, and science is insulated from outside criticism. Happy celebrations of progress on the part of the victors will not be met with any serious objection. This deflationary explanation of the appearance of progress is consistent with a relativist view of the changes between paradigms.

Kuhn also developed a second, very different account of the appearance of progress in science. This one seems to conflict with a relativist reading. Here Kuhn argued that science has a special kind of *efficiency,* and this efficiency results in a real form of progress across revolutions. The progress is measured in *problem-solving power;* the number and precision of solutions to problems in a scientific field tend to grow over time (1996, 170). It is hard to reconcile this claim with some of his discussions of incommensurability in earlier chapters. There he said that revolutions always involve losses as well as gains, and he also said that the standards that might be used to classify some problems as important and others as unimportant tend to change in revolutions. So we should be skeptical about whether the kind of measurement of problem solving power that Kuhn envisages in the last pages of *Structure* is compatible with the rest of the book.

If our later paradigms have more overall problem-solving power than our earlier ones, then it seems that we *are* entitled to regard the later ones as superior. This takes us away from relativism. Clearly Kuhn's aim was to work out an intermediate or moderate position (1996, 205–6). People will be arguing about this for a while to come.

So far I have been mostly discussing the comparison of different paradigms *within* science. What about the comparison of science with entirely different approaches to knowledge? Here Kuhn is sometimes read as a relativist, but this is straightforwardly a mistake. Kuhn thought that the overall structure of modern scientific investigation gives us a uniquely efficient way of studying the world. So if we want to compare scientific procedures of investigation with nonscientific ones, it is clear that Kuhn thought science was superior. He was not a relativist about this issue, and perhaps that is the most important issue.

That concludes my discussion of incommensurability and relativism. There is one more issue that is often grouped with the problem of incommensurability; this is the "theory-ladenness of observation." Kuhn argued that we cannot think of observation as a neutral source of information for choosing between theories, because what people see is influenced by their paradigm. Kuhn and a number of others developed radical views about observation around the same time. This is an important topic, as it challenges empiricism in a fundamental way. It will be discussed in chapter 10.

6.4 The X-Rated "Chapter X"

Kuhn's book starts out with his patient analysis of normal science. The middle chapters become more adventurous, and then the book climaxes with Chapter X. Here Kuhn puts forward his most radical claims. Not only do ideas, standards, and ways of seeing change when paradigms change; in some sense the *world* changes as well. Reality itself is paradigm-relative or paradigm-dependent. After a revolution, "scientists work in a different world" (1996, 135).

Philosophers and other commentators tend to split between two different attitudes toward this part of Kuhn's work. One group thinks that Kuhn exposes the fact that any idea of a single, stable world persisting through our various attempts to conceptualize it is an idea dependent on a failed view of science and outdated psychological theories. Kuhn, on this interpretation, shows that changing our view of science requires us to change our metaphysics too—our most basic views about reality and our relationship to it. Holding onto the idea of a single fixed world that science strives to describe is holding onto the last and most fundamental element of a conservative view of conceptual change.

That is one position. Others think that this whole side of Kuhn's work is a mess. When paradigms change, ideas change. Standards change also, and maybe the way we experience the world changes as well. But that is very different from claiming that the *world itself* depends on paradigms. The way we see things changes, but the world itself does not change.

I am in the second camp; the X-rated Chapter X is the worst material in Kuhn's great book. It would have been better if he had left this chapter in a taxi, in one of those famous mistakes that authors are prone to.

I should say immediately that it is not always clear how radical Kuhn wants to be in this chapter. Sometimes it seems that all he is saying is that our ideas and experience change. Also, there are some entirely reasonable claims we can make about changes to the world that result from paradigm changes. As paradigms change, scientists change their behavior and their experimental practices as well as their ideas. So some bits of the world change, in ordinary ways. And scientific revolutions result in new technologies that have far-reaching effects on the world we live in.

These changes can be far-reaching, but they are still restricted by the causal powers of human action. We can change plants and animals by controlled breeding and genetic engineering. We can dam rivers and also pollute them. But our reach is not indefinite. Kuhn discussed some cases in Chapter X that make it clear that he did not have these kinds of ordinary causal influences in mind. He discussed cases where changes in ideas about stars, planets, and comets led to astronomers "living in a different world," for example (1996, 117).

Most generally, though, the problem with these dramatic discussions is that Kuhn seems to think that the belief that we all inhabit a single world, existing independently of paradigms, also commits us to a naive set of ideas about perception and belief. But this is just not so. We might decide that perception is radically affected by beliefs and expectations, while still holding that perception is something that connects us to a single real world that we all inhabit.

Did Kuhn really make a mistake of this kind? As I have been so tough on him in this section, I should present my best "smoking gun" quote on the issue. Kuhn says:

> At the very least, as a result of discovering oxygen, Lavoisier saw nature differently. And in the absence of some recourse to that hypothetical fixed nature that he "saw differently," the principle of economy will urge us to say that after discovering oxygen Lavoisier worked in a different world. (1996, 118)

The passage is very strange. "Principle of economy"? Would it be *economical* for us to give up the idea that Lavoisier was living in the same world as the rest of us and acquiring new ideas about it? It is supposed to be *economical* to think that with every conceptual change of this kind, the scientist comes to live in a new, different world? Appeals to "economy" are often suspicious in the philosophy of science. They are usually weak arguments. This one also seems to have the accounting wrong.

From the point of view of a special kind of skeptical philosophical discussion, it can be considered "hypothetical" that there is a world beyond our momentary sensory experiences and ideas. But this is a very special sense of "hypothetical"! If we are trying to understand science as a social activity, as Kuhn is, there is nothing hypothetical about the idea that science takes place in a single, structured world that interacts with the community of scientists via the causal channels of perception and action.

The issues just discussed connect to another noteworthy feature of Kuhn's view of science. Kuhn opposed the idea that the large-scale history of science involves an accumulation of more and more knowledge about how the world really works. He was willing, on occasion, to recognize *some* kinds of accumulation of useful results as science moves along. There is an accumulation (maybe) of a kind of problem-solving power. But we cannot see, in science, an ongoing growth of knowledge about the structure of the world.

When Kuhn wrote about this issue, he often came back to cases in the history of physics. Like Popper and others, Kuhn seems to have been hugely influenced by the fall of the Newtonian picture of the world at the start of the twentieth century. Many philosophers of science seem to have been made permanently pessimistic about confirmation and the accumulation of factual knowledge by this episode. But Kuhn, and perhaps others, was surely *too* focused on the case of theoretical physics. He seems to have thought that we can *only* see science as achieving a growth of knowledge about the structure of the world if we can see this kind of progress in the parts of science that deal with *the most low-level and fundamental* entities and processes. But if we look at other parts of science—at chemistry and molecular biology, for example—it is much more reasonable to see a continuing growth (with some hiccups) in knowledge about how the world really works. We see a steady growth in knowledge about the structures of sugars, fats, proteins, and other important molecules, for example. There is no evidence that *these* kinds of results will come to be replaced, as opposed to extended, as science moves along. This type of work does not concern the most basic features of the universe, but it is undoubtedly science. It is possible that, when we try to work out how to describe the growth of knowledge over time in science, we should treat theoretical physics is a special case and *not* as a model for all science (McMullin 1984). So Kuhn's pessimism regarding the accumulation of knowledge about the structure of the world in science seems badly overstated.

6.5 Final Thoughts on Kuhn

Kuhn changed the philosophy of science by describing a tremendously vivid picture of scientific change. This picture was full of unexpected features, and Kuhn tried to shed light on traditional epistemological questions by looking at these questions from unusual angles. Most importantly, Kuhn attributed the success and the power of science to a delicate balance between factors in a complex and fragile mechanism. Science owes its strength to an interaction between the ordered cooperation and single-mindedness of normal science, together with the ability of these ordered behavioral patterns to break down and reconstitute themselves in revolutions. Periodic injections of disorder are just as essential to the process as the well-regulated behaviors found in normal science.

That is Kuhn's mechanism. Quite quickly, critics were able to find problems with this mechanism when interpreted as a description of how science actually works. I have already mentioned two important objections: single paradigms rarely have the kind of dominance that Kuhn describes, and large-scale changes can occur without crises. Many parts of Kuhn's mechanism are especially hard to apply to the history of biology, which Kuhn did not much discuss. Kuhn's account of the mechanisms behind scientific change is in several ways too tightly structured, too specific. The real story is more mixed. But Kuhn's was the first prominent attempt at a new *kind* of approach to the philosophy of science, a new kind of theory. These are theories that approach questions in the philosophy of science by looking at the social structure of science and the mechanisms underlying scientific change. This approach has flourished.

Back in the first chapter, I distinguished views that construe science broadly from those that construe it narrowly. Some philosophies of science are really extensions of more general theories in epistemology, psychology, or the philosophy of language. These views see the difference between science and everyday problem-solving as a matter of detail and degree. Kuhn's theory is nothing like this. His theory of science emphasizes the *differences* between science, narrowly construed, and various other kinds of empirical learning and problem-solving. Science is a form of organized behavior with a specific social structure, and science seems only to thrive in certain kinds of societies. As a consequence, science appears in this story as a rather fragile cultural achievement; subtle changes in the education, incentive structure, and political situation of scientists could result in the loss of the special mechanisms of change that Kuhn described.

Before moving on, as a kind of brief appendix I will mention some interesting connections between Kuhn's theory of science and a few other famous mechanisms for change. First, in some ways Kuhn's view of science has an "invisible hand" structure. The Scottish political and economic theorist Adam Smith argued in the *Wealth of Nations* ([1776] 1976) that individual selfishness in economic behavior leads to good outcomes for society as a whole. The market is an efficient distributor of goods to everyone, even though the people involved are each just out for themselves. Here we have an apparent mismatch between individual-level characteristics and the characteristics of the whole; selfishness at one level leads to the general benefit. But the mismatch is only apparent; it disappears when we look at the consequences of having a large number of individuals interacting together. We see something similar in Kuhn's theory of science: narrow-mindedness and dogmatism at the level of the individual lead to intellectual openness at the level of science as a whole. Anomaly and crisis produce such stresses in the normal scientist that an especially wholesale openness to novelty is found in revolutions. In the next chapter we will look at one critic who was suspicious of Kuhn on exactly this point; he thought Kuhn was trying to excuse and encourage the most narrow-minded and unimaginative trends in modern science.

Here is one more comparison, which is more complicated and requires more background knowledge. In the chapter on Popper, I briefly compared Popper's conjecture-and-refutation mechanism with a Darwinian mutation-and-selection mechanism in biology. A biological analogy can also be found in the case of Kuhn. During the 1970s, the biologists Stephen Jay Gould, Niles Eldredge, and others argued that the large-scale pattern seen in much biological evolution is "punctuated equilibrium" (Eldredge and Gould 1972). A lineage of organisms in evolutionary time will usually exhibit long periods of relative stasis, in which we see low-level tinkering but little change to fundamental structures. These periods of stasis or equilibrium are punctuated by occasional periods of much more rapid change in which new fundamental structures arise. (Note that "rapid" here means taking place in thousands of years rather than millions.) The rapid periods of change are disorderly and unpredictable when compared to the simplest kind of natural selection in large populations. The periods of stasis also feature a kind of "homeostasis" in which the genetic system in the population tends to *resist* substantial change.

The analogy with Kuhn's theory of science is striking. We have the same long periods of stability and resistance to change, punctuated by unpredictable, rapid change to fundamentals.

The theory of punctuated equilibrium in biology was controversial for a time, especially because it was sometimes presented by Gould in rather radical forms (Gould 1980). The idea of a kind of "homeostatic" resistance to change brought about by the genetic system is a tendentious idea, for example. And the idea that ordinary processes of natural selection do not operate normally during the periods of rapid change, but are replaced by other kinds of processes, is also very unorthodox. But as the years have passed, the idea of punctuated equilibrium has been moderated and has passed, in its more moderate form, into mainstream biology's description of at least some patterns in evolution (Futuyma 1998).

Gould also wrote a paper called "Eternal Metaphors in Paleontology" (1977), in which he argued that the history of theorizing about the history of life sees the same basic kinds of ideas about change come up again and again, often mixed and matched into new combinations. The analogy between Kuhn's theory and the biological theory of punctuated equilibrium shows the same kind of convergence on a story about processes of change. We also see similar kinds of opposition between the "punctuated equilibrium" story found in Kuhn and Gould, on one side, and a more uniform, one-process view of change found in Popper and in some rival views of biological evolution, on another. I say "convergence" here, but Gould has acknowledged the influence of Kuhn's picture of science on him when he was working out his biological ideas in the 1960s and 1970s (Gould 2002, 967). Kuhn himself was also interested in (different) possible analogies between his picture of science and evolutionary change (1996, 171–72).

Further Reading

The readings from chapter 5 are relevant here as well. For a very detailed discussion of Kuhn's philosophy, see Hoyningen-Huene 1993. Kitcher, *The Advancement of Science* (1993), contains thorough (and sometimes difficult) critical discussions of some of Kuhn's most influential arguments about revolutions and progress, including several aspects of the incommensurability problem. Kitcher also reanalyzes several of Kuhn's big historical examples. Doppelt 1978 is a clear discussion of incommensurability of standards.

7

Lakatos, Laudan, Feyerabend, and Frameworks

7.1 After *Structure*

The period after the publication of Kuhn's book was one of intense and sometimes heated discussion in all the fields that try to understand science. In this chapter we will discuss some other philosophical accounts of science developed around this time, all of them developed in interaction with Kuhn or in response to him.

Then we will pause for a breath, and we'll consider some general patterns in the ideas described in the previous chapters.

First, we will look at the views of Imre Lakatos. Lakatos's main contribution was the idea of a *research program*. A research program is similar to a paradigm in Kuhn's (broad) sense, but it has a key difference: we expect to find more than one research program in a scientific field at any given time. The large-scale processes of scientific change should be understood as *competition between research programs*.

It should be obvious from the previous chapters that this was an idea waiting to be developed. Kuhn's insistence that scientific fields usually have only one paradigm operating at any time was criticized right from the initial publication of *Structure*. Lakatos was the first person to develop a picture of science in which larger paradigm-like units operate in parallel and compete in an ongoing way. Lakatos's own development of this idea has problems, and it was embedded within a general philosophical program that has very peculiar features. Shortly after Lakatos's work, another philosopher, Larry Laudan, worked out a superior version of the same basic idea.

After looking at Lakatos and Laudan, we turn to Paul Feyerabend, the wild man of twentieth-century philosophy of science.

Feyerabend is the most controversial and extreme figure contributing to the debates discussed in this book. I called him "the" wild man, even though there have been various other wild men—and wild women—in the field besides Feyerabend. But Feyerabend's voice in the debates was uniquely wild. He argued for "epistemological anarchism," a view in which rules of method and normal scientific behavior were to be replaced by a freewheeling attitude in which "anything goes."

Kuhn, Lakatos, and Feyerabend all interacted and developed some of their ideas in response to each other (with the exception perhaps of much influence of Lakatos on Kuhn). Feyerabend's most important work (1975) was written, he said, as a kind of letter to Lakatos, but Lakatos died in 1974 before writing a reply.

7.2 Lakatos and Research Programs

Imre Lakatos had a remarkable life. Born in Hungary, he was a member of the resistance to Nazi occupation during World War II. After the war he worked in politics and was jailed for over three years by the Stalinist regime. He left Hungary, made his way to England, and eventually ended up at the London School of Economics working with Popper. Lakatos often claimed that his main ideas about science were implicit in Popper or were one side of Popper's views. Although there is some truth in this, it is better to consider Lakatos's ideas in their own terms.

Lakatos's reaction to Kuhn's work was one of dismay. He saw Kuhn's influence as *destructive*—destructive of reason and ultimately dangerous to society. For Lakatos, Kuhn had presented scientific change as a fundamentally irrational process, a matter of "mob psychology" (1970, 178), a process where the loudest, most energetic, and most numerous voices would prevail regardless of reasons. The interpretation I gave of Kuhn in the previous two chapters is very different; in this interpretation Kuhn saw science as an almost miraculously well-structured machine for exploring the world. Even the disordered episodes found in revolutions have a positive role in the functioning of the whole. Lakatos, in contrast, saw the disorder in Kuhn's picture as no more than dangerous chaos. But Lakatos also saw the force of Kuhn's historical arguments. So his project was to rescue the rationality of science from the damage Kuhn had done.

Lakatos had some views about the relation between the history of science and the philosophy of science that are spectacularly strange. Lakatos argued that historical case studies should be used to assess philosophical views of science. Fine, so far. But he also said that we should write "rational reconstructions" of the historical episodes, in which scientists' decisions are made to look as rational as possible. We should then separately (or in footnotes) point out places where the rational reconstruction is not an accurate description of what actually went on. So it is OK to *deliberately misrepresent* what happened in the past, so long as the footnotes set things straight (see Lakatos 1970, 138 n, 140 n). What matters most is that in the main discussion we are able to spin a story in which the scientific decisions come out looking rational. I have never understood why this idea is not met with more amazement and criticism from philosophers. (Hacking 1983 is a vigorous exception.)

In among all this, however, Lakatos developed a view of the organization of science that was very influential. This is known as his *methodology of scientific research programs* (though he spelled it "programmes," in the British way).

A research program, for Lakatos, is roughly analogous to a Kuhnian paradigm (in the broad sense). The big difference, as I said above, is that there is usually more than one research program per field at any given time. According to Lakatos, competition between research programs is what we actually find in science, and it is also essential to rationality and progress. This view was applied to all of science, from physics to the social sciences.

A research program is a historical entity; it evolves over time. It will contain a sequence of related theories. Later theories are developed in response to problems with the earlier ones. For Lakatos, as for Kuhn, it is common and justifiable for a research program to live for a while despite empirical anomalies and other problems. Workers within a research program typically have some commitment to the program; they do not reject the basic ideas of the program as soon as something goes wrong. Rather, they try to modify their theories to deal with the problem. However, for Lakatos as for Kuhn, research programs are sometimes abandoned. So a complete theory of scientific change must consider two different kinds of change: (1) change *within* individual research programs, and (2) change at the level of the *collection* of research programs within a scientific field.

A research program has two main components, in Lakatos's view. First, it contains a *hard core*. This is a set of basic ideas that are essential to the research program. Second, a research program contains a *protective belt*. This is a set of less fundamental ideas that are used to apply the hard core to actual phenomena. The detailed, specific versions of a scientific theory that can actually be tested will contain ideas from the hard core combined with ideas from the protective belt.

For example, the Newtonian research program of eighteenth-century physics has Newton's three laws of motion and his gravitational law as its hard core. The protective belt of Newtonianism will change with time, and at any time it will contain detailed ideas about matter, a view about the structure of the universe, and mathematical tools used to link the hard core to real phenomena. The nineteenth-century Darwinian research program in biology has a hard core that claims that different biological species are linked by descent and form a family tree (or perhaps a very small number of separate trees). Changes in biological species are due mostly to the accumulation of tiny variations favored by natural selection, with some other causes of evolution playing a secondary role. The protective belt of nineteenth-century Darwinism is made up of a shifting set of more detailed ideas about which species are closely related to which; ideas about inheritance, variation, competition, and natural selection; ideas about the distribution of organisms around the earth; and so on.

We now reach Lakatos's principles of scientific change. Let us first look at change *within* research programs. The first rule is that changes should only be made to the protective belt, never to the hard core. The second rule is that changes to the protective belt should be *progressive*. Here Lakatos borrowed from Popper's ideas. A progressive research program constantly expands its application to a larger and larger set of cases, or strives for a more precise treatment of the cases it presently covers. A progressive research program is one that is succeeding in increasing its predictive power. In contrast, a research program is *degenerating* if the changes that are being made to it only serve to cover existing problems and do not successfully extend the research program to new cases. Lakatos assumed, like Kuhn, that all research programs are faced with anomalies, unsolved empirical problems, at any time. A degenerating research program is one that is falling behind, or only barely keeping up, in its attempt to deal with anomalies. A progressive research program fends off refutation and *also* extends itself to cover new phenomena. Lakatos thought that, in principle, we could measure how quickly a research program is progressing.

Now let us look at the higher level of change in Lakatos's system, change at the level of the collection of research programs present in a scientific field.

Each field will have a collection of research programs at any given time, some of which are progressing rapidly, others progressing slowly, and others degenerating. You might be thinking that the next rule for Lakatos is obvious: "choose the most progressive research program." That would establish a decision procedure for scientists looking at their whole field, and it would give us a way of deciding who is making rational or irrational decisions. But that is not what Lakatos said.

For Lakatos, it is acceptable to *protect* a research program for a while, during a period when it is degenerating. It might recover. This is even the case when another research program has overtaken it (Lakatos 1971). The history of science contains cases of research programs recovering from temporary bad periods. So a reasonable person can wait around and hope for a recovery. How long is it reasonable to wait? Lakatos does not say.

Feyerabend swooped on this point (1975). For him it was the Achilles' heel in Lakatos's whole story. If Lakatos does not give us a rule for when a rational scientist should give up on one research program and switch to another, his account of rational theory choice is completely empty.

So *is* there a third rule that tells us how to handle decisions between research programs? Not really. Lakatos did say that the decision to stay with a degenerating research program is a *high-risk* one (1971). So Lakatos might advise the rational scientist to stay with a degenerating research program only if he or she is willing to tolerate a high-risk situation. And Lakatos is right that different people can reasonably have very different attitudes toward risk. But Lakatos did not follow the suggestion up, to close the gap in his theory. The tremendous *appearance* of order and methodological strictness in Lakatos's philosophy of science is much undermined by his failure to say something definite about this crucial point. Feyerabend was right to see a mismatch between the rhetoric and the reality of Lakatos's views. Indeed, it sometimes looks as if the whole point of Lakatos's project was to give us a way of *retrospectively* describing episodes in science as rational.

This is a good place to emphasize the vast difference between Lakatos and Kuhn in their underlying attitudes. Kuhn has a deep trust in the shared standards implicit in paradigms and the ability of science to find a way forward after crises, despite some groping and flailing along the way. For Kuhn, once we rid our picture of science of some myths, the picture we are left with is fundamentally healthy; Kuhn *trusts* science left in the hands of implicit shared values. Lakatos, in contrast, wants to have the whole enterprise guided by methodological rules—or at least he needs for us to be able to tell ourselves a *story* of that kind.

Let us not worry further about the oddities in Lakatos's view. Instead, we can ask, Does his picture of the structure of science have any useful elements? Once we ask this, I think it is clear that the basic idea of competing research programs is a useful one. Certainly there are some fields where this seems a far more accurate description of what goes on than Kuhn's paradigm-based view. Psychology is an obvious example. Current work in "evolutionary psychology" looks a lot like a research program in something like Lakatos's sense (Barkow, Cosmides, and Tooby 1992).

We might also consider the possibility of *mixtures* of Kuhn-like and Lakatos-like stories. In biology, what we often find is consensus about very basic principles but competition between research programs at a slightly lower level. Looked at very broadly, evolutionary biology might contain something close to a single paradigm: the "synthetic theory," a combination of Darwinism and genetics. But at a lower level of generality, we seem to find competing research programs. The "neutral theory of molecular evolution" is a research program that tries to understand most variation and change at the molecular genetic level in terms of random processes rather than natural selection (Kimura 1983). This research program is compatible with the central claims of the synthetic theory, but it conflicts with some standard ways in which the synthetic theory is applied to genetic variation within populations.

In the case of the neutral theory and also in other cases, what we find is a research program being "budded off" from the mainstream of biology and being explored for a few decades to see how much it can explain. Then it may turn out that the limits of the research program are reached, at which point it is moderated and folded back into the mainstream. That is what seems to be happening with the neutral theory.

So we now have the tools for describing a range of different large-scale processes in science. Some fields may have dominant paradigms and Kuhnian normal science. Others may have competing research programs. Some might have very general paradigms plus lower-level research programs budding off periodically. (Here I should note that the term "research program" can also sometimes be used to describe different *non*competing approaches within a single field.)

I have been discussing the usefulness of the research program idea in describing how science *actually* works. There is also the possibility of *normative* theories that make use of this concept. But I will not follow up that idea further within Lakatos's framework. We will return to the topic in the next section.

I will make one last point before leaving Lakatos. In the last few chapters, I have contrasted Popper, who called for permanent open-mindedness and criticism, with Kuhn, who endorsed a tenacious commitment to the basic ideas of a paradigm. This is a standard way of marking a fundamental disagreement between Kuhn and Popper. There is a *bit* more complexity here, however. Lakatos, as I said, saw many of his ideas as implicit in Popper. And we can indeed find passages in Popper where he accepts that theories should not be discarded at the first sign of trouble but, instead, should be protected initially to see if they can overcome their problems (Popper 1963, 49; 1970, 55). So is there really no difference, or much less difference, between Popper and Kuhn on this point? Has Popper retreated from one of his most basic ideas? Not really. When Popper came to directly confront Kuhn's arguments on this issue (1970), he did *not* choose to blur the difference between his view and Kuhn's. He said that Kuhnian normal science does sometimes exist, but it was not nearly as common as Kuhn said. And more importantly, he regarded it as a bad thing, which should not be encouraged.

7.3 Laudan and Research Traditions

In an interesting book called *Progress and Its Problems* (1977), Larry Laudan developed a view that is similar to Lakatos's in basic structure but which is far superior. Like Lakatos, Laudan thought that Kuhn had described science as an irrational process, as a process in which scientific decision-making is "basically a political and propagandistic affair" (1977, 4). This reading of Kuhn (I say yet again) is inaccurate. But Laudan also recognized the power of Kuhn's discussions of historical cases. Like Lakatos, Laudan wanted to develop a view in which paradigm-like entities could coexist and compete in a scientific field. He gave many cases from the history of science to motivate this picture. So we are heading toward the idea of a research program. But in an understandable piece of product differentiation, Laudan called the large-scale units of scientific work "research traditions" rather than research programs.

The difference between Laudan and Lakatos is not just terminological. Laudan's description of research traditions makes more sense than Lakatos's account. Lakatos saw the sequence of theories within a research program as linked very closely by logic; each new theory was supposed to have a broader domain of application than its predecessor in that research program. And for Lakatos, the hard core never changes. For Laudan, the theories grouped within research traditions are more loosely related. There can be some movement of ideas in and out of the hard core. Moreover, for Laudan there is nothing unusual or bad about a later theory covering less territory than an earlier one; sometimes a retreat is necessary. For Laudan, theories can also break away from one research tradition and be absorbed by others. For example, the early thermodynamic ideas of Sadi Carnot were developed within a research tradition that saw heat as a fluid ("caloric"), but these ideas were taken over in time by a rival research tradition that saw heat as the motion of matter.

Another key innovation in Laudan's account is his distinction between the *acceptance* and the *pursuit* of theories.

The philosophies discussed here so far have tended to recognize just *one kind* of attitude that scientists can have to theories. Usually the attitude of a scientist to a theory has been treated as something like *belief*. Of course, belief can come in degrees; there are cautiously held and firmly held beliefs.

Scientists will often—and should often—have only a cautious belief. But still the idea here is that there is one basic *kind* of attitude—belief or something like it—though it comes in degrees. Laudan argued that there are *two* different kinds of attitudes to theories and research traditions found in science, *acceptance* and *pursuit.* Acceptance is close to belief; to accept something is to treat it as true. But pursuit is different. It involves deciding to work with an idea, and explore it, for reasons *other* than confidence that the idea is likely to be true. Crucially, it can be reasonable to *pursue* an idea that one definitely does not *accept.* Someone might have reason to believe that *if* the idea was true, it would be of huge importance and the payoff from working on it would be high. Someone might think that although an idea is not likely to be true, it should be explored, and that she or he is the person best equipped to do so. There is a whole constellation of different reasons that a person might have for working with a scientific idea.

Laudan built the distinction between acceptance and pursuit into his account of rational decision-making in science. He was able to give some fairly sharp rules where Lakatos had not. For Laudan, it is always rational to *pursue* the research tradition that has the highest current *rate of progress* in problem-solving (1977, 111). But that does not mean one should *accept* the basic ideas of that research tradition. The acceptability of theories and ideas is measured by their present overall *level* of problem-solving power, not by the rate of change. We should accept (perhaps cautiously) the theories that have the highest level of problem-solving power. So a scientist might be inclined to accept the ideas in a mainstream research tradition but work on a more marginal research tradition that has a spectacular rate of progress. For Laudan that decision would be a rational one.

With any rule like this, it will be possible to think of cases where the rule might lead a person astray. What if a research tradition has a low rate of progress right now, but there is good reason to think it might take off very soon? This is the kind of possibility that made Lakatos hesitate. Laudan clearly hoped that the distinction between acceptance and pursuit would help with this kind of problem, and so it does. But he does not try to lay down rules that will deal with every possible situation, including all the various kinds of bad luck. Here we run into a general problem about the aims of philosophy of science; it is very unclear what *kinds* of principles a philosophy of science should be looking for. Some think that looking for rules of procedure, even sophisticated ones like Laudan's, is just a mistake. But it is fair to say that Laudan was able to give quite an impressive normative theory of science using the idea of competition between research traditions.

Laudan's theory was impressive, but here is an interesting gap in both theories discussed in this chapter that some readers may have picked up on already. Both Lakatos and Laudan were interested in the situation where a scientist is looking out over a range of research programs in a field and deciding which one to join. But here is a question that neither of them seemed to ask: does the answer depend on *how many* people are *already* working in a given research program? Both Lakatos and Laudan seemed to think that it would be fine for their theories to direct everyone to work on the same research program, if it was far superior to the others. But perhaps that is a mistake. Science might be better served by some kind of mechanism in which the field *hedges its bets*. That suggests a whole different question that might be addressed by the philosophy of science: what is the best *distribution* of workers across a range of research programs?

There are two different ways of approaching this new question. One way is to look at individual choices. Does it make sense for *me* to work on research program 1 rather than research program 2, given the way people are already distributed across the two programs? Is research program 1 overcrowded? Perhaps Lakatos and Laudan thought this question was not relevant to their project because it seems to require introducing selfish goals into the picture. But we can also approach the issue another way. We can ask, Which distribution of people across rival research programs is *best for science*?

Kuhn, interestingly, was aware of this issue, especially in his work after *Structure* (e.g., 1970). This is ironic because Kuhn did not think that ongoing competition between paradigms was usually found in science. But Kuhn did say that one of the strengths of science lay in its ability to *distribute risk* by having different scientists make different choices, especially during crises. Lakatos and Laudan were in a good position to make a really thorough investigation of this issue, but they did not (see also Musgrave 1976). It was not until more recently that this question was brought into sharp philosophical focus. We will take up the issue in detail in chapter 11.

7.4 Anything Goes

Now we turn to Paul Feyerabend, the most controversial and adventurous figure in the post-Kuhn debates. Feyerabend, like many key figures in this book, was born in Austria. He fought in the German infantry during World War II and was wounded. He switched from science to philosophy after the war and eventually made his way to the University of California at Berkeley, where he taught for most of his career. Feyerabend was initially influenced by Popper, with whom he worked for a year in the 1950s. But by the early 1960s, he was moving toward the adventurous views for which he became famous. He and Kuhn influenced each other significantly. (In discussing Feyerabend after Lakatos and Laudan, I am departing from the chronological organization of this book).

So what were his notorious ideas? A two-word summary gets us started: *anything goes*. Feyerabend's most famous work was his 1975 book *Against Method*. Here he argued for "epistemological anarchism." The epistemological anarchist is opposed to all systems of rules and constraints in science. Great scientists are *opportunistic* and *creative*, willing to make use of any available technique for discovery and persuasion. Any attempt to establish rules of method in science will result only in a straitjacketing of this creativity. We see this, Feyerabend said, when we look at the history of science. Great scientists have always been willing to break even the most basic methodological rules that philosophers might try to lay down. The only rule that we can be *sure* will not impede imagination and progress is this: anything goes.

In defending his position, Feyerabend made use of a range of ideas about scientific language and the psychology of observation. Like Kuhn, he thought that rival scientific theories are often linguistically incommensurable (section 6.3 above). He argued that observations in science are contaminated with theoretical assumptions and hence cannot be considered a neutral test of theory. These arguments were based on speculative ideas about scientific language, and they are not very convincing. His more interesting arguments are of two kinds. These involve both the history of science and a direct confrontation of some hard questions about how science relates to freedom and human well-being.

Before launching into this unruly menagerie of ideas, we need to keep in mind a warning that Feyerabend gave at the start of *Against Method*. He said that the reader should *not* interpret the arguments in the book as expressing Feyerabend's "deep convictions." Instead, they "merely show how easy it is to lead people by the nose in a rational way" (1975, 32). The epistemological anarchist is like an "undercover agent" who uses reason in order to destabilize it. Again we are being told by an author not to trust what we are reading. It is hard to know what to make of this, but I think it is possible to sort through Feyerabend's claims and distinguish some that do represent his "deep convictions." Feyerabend's deepest conviction was that *science is an aspect of human creativity*. Scientific ideas and scientific change are to be assessed in those terms.

In his article for the *Routledge Encyclopedia of Philosophy* (1998), Michael Williams suggests that we think of Feyerabend as a late representative of an old skeptical tradition, represented by Sextus Empiricus and Montaigne, in which the skeptic "explores and counterposes all manner of competing ideas without regarding any as definitely established." This is a useful comparison, but it is only part of the story. To capture the other part, we might compare Feyerabend to Oscar Wilde, the nineteenth-century Irish playwright, novelist, and poet who was imprisoned in England for homosexual behavior. Wilde is someone who liked to express strange, paradoxical claims about knowledge and ideas ("I can believe anything so long as it's incredible"). But behind the paradoxes there was a definite message. For Wilde, the most important kind of assessment of ideas is *aesthetic* assessment. A book or an idea might look immoral or blasphemous, but if it is beautiful, then it is worthwhile. Other standards—moral, religious, logical—should never be allowed to get in the way of the free development of art. This, I suggest, is close to Feyerabend's view; what is important in all intellectual work, including science, is the free development of creativity and imagination. Nothing should be allowed to interfere with this.

Feyerabend's focus on values and creativity guided his readings of others. His paper "Consolations for the Specialist" (1970) shows him to be one of the most perceptive critics of Kuhn. Most philosophers of science found an alarming disorder in Kuhn's view of science. Feyerabend found the opposite: an *incitement* for scientists to become orderly and mechanical. Feyerabend saw Kuhn as glorifying the mind-numbing routine of normal science and the rigid education that Kuhn thought produced a good normal scientist. He saw Kuhn as encouraging the worst trends in twentieth-century science toward professionalization, narrow-mindedness, and exclusion of unorthodox ideas.

Feyerabend recognized the "invisible hand" side of Kuhn's story, his attempt to argue that individual narrow-mindedness is all for the best in science. Back in chapter 5, I said that it is often hard to distinguish the descriptive from the normative in Kuhn's discussions. Feyerabend saw this ambiguity in Kuhn's writing as a *deliberate rhetorical device* for insinuating into the reader a positive picture of the most mundane type of science. For Feyerabend, the mind-set that Kuhn encouraged also leads to a lack of concern for the moral consequences of scientific work.

Feyerabend also argued that Kuhn was factually wrong about the role of normal science in history. According to Feyerabend, paradigms almost never succeed in exerting the kind of control Kuhn described. There are always imaginative individuals trying out new ideas.

Feyerabend was not, as he is sometimes portrayed, an "enemy of science." He was an enemy of *some kinds* of science. In the seventeenth century, according to Feyerabend, science was the friend of freedom and creativity and was heroically opposed to the stultifying grip of the Catholic church. He admired the scientific adventurers of this period, especially Galileo. But the science of Galileo is not the science of today. Science, for

Feyerabend, has gone from being an ally of freedom to being an enemy. Scientists are turning into "human ants," entirely unable to think outside of their training (1975, 188). And the dominance of science in society threatens to turn man into a "miserable, unfriendly, self-righteous mechanism without charm or humour" (175). In the closing pages of *Against Method,* he declares that society now has to be freed from the strangling hold of a domineering scientific establishment, just as it once had to be freed from the grip of the One True Religion.

7.5 An Argument from History That Haunts Philosophy

Let us now look at the argument that is perhaps most central to Feyerabend's work. This is an argument from history.

A large part of *Against Method* is taken up with a discussion of Galileo's arguments against his Aristotelian opponents in the early seventeenth century (see section 1.5). Galileo aimed to defend the literal truth of Copernicus's claim that the earth goes round the sun rather than vice versa. One of the things Galileo had to confront was a set of obvious arguments from experience against a moving earth. For example, when a ball is dropped from a tower, it lands at the foot of the tower even though, on Copernicus's view, the tower has moved a significant distance (along a huge circle) while the ball is in the air. All of our everyday experience of motion suggests that the earth is stationary. These are not arguments from the wisdom of Aristotle or the sayings of the biblical apostles; they are arguments from what we observe every day. If empiricism in philosophy has any teeth at all, Feyerabend claims, it entails that people in the seventeenth century had excellent reasons to resist Galileo and believe that the earth is not moving.

Galileo, of course, rejected the arguments. In his *Dialogue concerning the Two Chief World Systems* ([1632] 1967), he patiently tries to show that Copernicus's model is compatible with everyday experiences of motion. If the earth is moving, then a ball dropped from a tower has a *mixed* kind of motion. It is falling toward the earth but is also moving in a huge circle just as the tower is. Our everyday perception of motion is unable to distinguish the case where both tower and ball have a circular motion from the case where neither does.

Galileo defuses the arguments, but he does not suggest that this is easy. Indeed, he marvels at Copernicus and others who have "through sheer force of intellect done such violence to their own senses as to prefer what reason told them over that which sensible experience plainly showed them to be the contrary" (quoted in Feyerabend 1975, 101).

According to Feyerabend, what Galileo had to do was to *create* a different kind of observational description of the world, one in which descriptions of apparent motion were compatible with the Copernican hypothesis. Only then would the arguments for Copernicanism become plausible. What science had to do, here and in other cases, was to *break through* the constraints of an outdated worldview that had permeated even the most basic observational description. Science, for Feyerabend, is often a matter of *challenging rather than following* the lessons of observation.

What we see here is a simple case of something that Feyerabend regards as ubiquitous. A very basic empiricist principle, of the kind dear to philosophers, would seem to be pointing people in the seventeenth century *away* from the scientific theory that we think now to be true. The philosopher complacently spouts generalities about how science is great because it is responsive to observational data. But history suggests that the principles the philosopher likes so much would *steer us in the wrong direction* if people back in the crucial period had applied them.

Through all the exaggerations, deliberate provocations, jokes, insults, and outrageous statements in Feyerabend's works, this form of argument runs as a constant and challenging thread. Are there any principles of method, measures of confirmation, or summaries of the scientific strategy that do not *fail* the great test of the early seventeenth century? Look at the massiveness of the rethinking that Galileo urged, and the great weight of ordinary experience telling against him. Given these, would all traditional philosophical accounts of how science works, especially empiricist accounts, have instructed us to stick with the Aristotelians rather than take a bet on Galileo? This is the Feyerabendian argument that haunts philosophy of science.

However, Feyerabend massively overextends his argument, into a principle that cannot be defended: "Hence it is advisable to let one's inclinations go against reason *in any circumstances*, for science may profit from it" (1975, 156). Feyerabend claims that because some principle or rule *may* go wrong, we should completely ignore it. The claim is obviously crazy. The policy of catching trains that are scheduled to take you where you plan to go is a policy that *may* go wrong. The train might crash. Or if you caught a different train, you might meet the love of your life on the way. All that is possible, but no rational person regards these mere possibilities as sufficient to discredit the rule that it is best to take trains that are scheduled to go where you want to go. All applications of everyday rules of rational behavior presuppose judgments about which outcomes are probable, or typical, and which are far-fetched or unlikely. Sometimes we might be able to put exact numbers on the chances of these different outcomes, but often we cannot. Sometimes we rely on informal judgment to guide us. In rational behavior, nothing is guaranteed, but some policies and rules can be justified despite this. (The modern Bayesian theory of belief and action, which will be discussed in chapter 14, is based upon this idea.) And what is true for everyday behavior is true also for science. The mere *possibility* that a rule might lead to bad consequences proves very little. We need more than a mere possibility before we have grounds to doubt a principle. Science "may profit" from all kinds of strange decisions, but "may" is not enough.

So we have a mixture of good and bad argument in Feyerabend's treatment of these issues. It would be foolish for a philosopher to ignore the strangeness of having one's favorite principles of theory choice come out in favor of the seventeenth-century Aristotelians and against Galileo. The example is *so* important in the history of science that it takes a brave philosopher to ignore it. It is by no means clear that well-developed philosophical theories do make the wrong call, though. Recall Laudan's rules discussed in section 7.4 above; they would probably steer us rather well here. These rules would at *least* tell us to *pursue* the Galilean program, on account of its rate of increase in problem-solving power. And in time, it would become rational to accept the Galilean view.

7.6 Pluralism and the Ramblings of Madmen

For Feyerabend, science benefits from the presence of a range of alternative ideas and perspectives. Let us look at Feyerabend's ideas about pluralism and diversity in science.

In his 1970 paper about Kuhn and Lakatos, Feyerabend proposed two general principles that should guide science. We should keep in mind that "rules are made to be broken" in Feyerabend's view. But the principles are worth discussing.

The first rule Feyerabend called the "principle of tenacity." This principle tells us to hold onto attractive theories despite initial problems and allow them a chance to develop their potential.

That is a start, but if everyone followed this rule, nothing would ever change. So Feyerabend adds a second principle, the "principle of proliferation." This principle tells us to make up new theories, propose new ideas.

Kuhn says the proliferation of new ideas should wait for a crisis. But why not put forth new ideas all the time? Thus we reach Feyerabend's ideal picture of science: we have a population of people happily developing their theories and also trying to think up new ones. Some pedestrian work is needed to help develop the existing ideas, but this should not interfere with the imaginative work.

Feyerabend claimed to be following the tradition of the philosopher and political theorist John Stuart Mill. In his classic paper *On Liberty* ([1859] 1978), Mill argued that society benefits from a diversity of ideas and ways of life. The constant proliferation of new theories creates a "marketplace of ideas," in which many options are explored and the best prevail. Some of Feyerabend's descriptions of ways in which science might benefit from unusual sources of ideas—"the ramblings of madmen" (1975, 68)—were intended as descriptions of *inputs* into this marketplace. (See also Lloyd 1997.) Feyerabend argued that we can often only perceive the limitations of our current perspective if we try to step outside it, at least temporarily. Achieving a novel, external perspective on ideas that we usually assume uncritically is often the beginning of progress. What we usually consider "established facts," including observable facts, are often laden and contaminated with prejudices and outdated ideas. Any source of an external vantage point is to be valued. Alternative theories, even theories with massive problems, can provide this kind of external vantage point.

These ideas about the need for external vantage points in challenging familiar assumptions are interesting. And the idea of a marketplace of ideas is a powerful one. But Feyerabend's account seems to have left out something very important. And this omission undermines his version of the "marketplace of ideas" doctrine.

What is missing in Feyerabend's picture is some rule or mechanism for the *rejection* and *elimination* of ideas. Feyerabend gives a recipe that, if it was followed, would lead to the accumulation of an ever-increasing range of scientific ideas being discussed in every field. Some ideas would probably become boring and might be dropped for that reason. But aside from that, there is *no* way for an idea to be taken *off* the table. So a question immediately becomes pressing: what are we supposed to do when we have to apply one of these theories to a practical problem? What do we do when the bridge has to be built? Which ideas should we use? Not the most "creative" ones, surely! Feyerabend never gave a satisfactory answer to this question.

If what we really want from science is to have a community full of lively, imaginative discussion, then Feyerabend's recipe for scientific behavior is an appropriate one. Science then would be very similar to art. But if part of the role of science is to guide us in solving practical problems, then Feyerabend's recipe seems completely misguided. If science must be applied to problems, there must be a mechanism of *selection* in science, a mechanism for the *rejection* of some ideas. Proliferation of alternatives is *part* of science, and another part is selection among alternatives.

In the last few years, for example, the government of Thabo Mbeki in South Africa has shown an interest in radical ideas about the causation of AIDS. According to these ideas, the virus identified by mainstream science as the cause of AIDS, HIV, is regarded as either relatively unimportant or altogether harmless. In reply to the storm of criticism that resulted, Mbeki has sometimes said that he is simply interested in an open-minded questioning of theories and the exploration of diverse possibilities. Surely that is a properly scientific attitude? This reply has been rightly criticized as disingenuous. Science needs the invention of alternatives, but it also needs mechanisms for pruning the range of options and abandoning some. When the time comes to apply scientific ideas in a public health context, this selection process is of paramount importance. Then we must *take from* science the well-supported view that AIDS is caused by a virus transmitted through body fluids, and we must guide policy and behavior with this view.

Michael Williams, in the encyclopedia article I mentioned earlier, says the following about how Feyerabend looks in retrospect: "While some of his views may strike some philosophers as overstated, their general spirit has some claim to be seen as today's conventional wisdom." I would give an almost opposite summary. While a few individual pieces of Feyerabend's view have become something close to conventional, the "general spirit" of his work involves a principle that is still unconventional because it is clearly false. This is the principle that we should think of the social role of science in the same way we think of the social role of art. On the contrary, imagination and creativity are one side of science but not the only side.

7.7 Taking Stock: Frameworks and Two-Process Theories of Science

In this section I will discuss a general theme that has appeared several times over the last few chapters. This section is a "time out" from the chronological story central to the book.

The theme I will discuss has reverberated continually through twentieth-century philosophy and continues to be important. I will introduce it as a distinction between views about scientific change, or conceptual change more generally. The distinction is between *one-process* and *two-process* theories. If we are trying to understand scientific change, should we recognize two "levels" in science, with different types of change occurring at these two levels? More precisely, should we see scientific change as involving (1) changes made *within* the boundaries provided by a general framework and (2) changes at the level of the frameworks themselves?

The alternative (or rather, one alternative) is to give a unified account, in which there is no qualitative distinction made between two levels, layers, or kinds of change.

The opposition between Popper and Kuhn can be used to make the contrast vivid. For Popper, scientific change always involves the same process—

the cycle of conjecture and refutation. This is what we see at the level of small changes to details and also at the level of fundamental changes in worldview. For Kuhn, in contrast, there are two qualitatively different kinds of change in science; changes within paradigms and change between paradigms involve fundamentally different processes. A paradigm (in the broad sense) is a clear example of a framework in the sense relevant here. Change within a framework is *guided* by principles supplied by the framework. Because frameworks supply these principles, moves *between* frameworks are more problematic, hard to describe, and often disorderly.

Let us move to another contrast that is fairly clear and vivid: the contrast between Carnap and Quine. Carnap, in his later philosophy, used the term *linguistic framework* and distinguished moves made within these frameworks from changes made between them. The principles that are fundamental to a framework will appear as analytic sentences if they are explicitly stated. Moves made within a framework involve the assessment and testing of synthetic claims. For Carnap, many alternative frameworks are possible, and people can switch between them. These switches, however, involve a different kind of process from moves made within them. Moves made between frameworks are not sensitive to *particular* factual results; they are sensitive to a kind of pragmatic assessment of the *overall* usefulness of the framework. If one framework seems not to be working, then we try another. Carnap's frameworks are "thinner" than Kuhn's; they involve just basic linguistic and logical rules, not scientific principles.

Quine, in "Two Dogmas of Empiricism" and elsewhere, argued against this two-layered view of language and knowledge. For Quine, all changes made to our belief system, whether large or small, involve the same kind of holistic tinkering with the web of belief. We accommodate experience by making as *few* changes as possible and keeping our worldview as *simple* as we can. There is no distinction between changes within and changes between frameworks.

How might we decide between a one-process view and a two-process view? Within twentieth-century philosophy, many people were persuaded by Quine's holism. These arguments were based on very general considerations and not on the history of specific episodes in science. Quine's most powerful argument is usually seen to be his claim that there is no way to mark out the distinction between changes within and changes between frameworks in a way that is scientific and does not beg the question.

Kuhn, however, had no problem distinguishing normal science from revolutionary change in actual scientific cases. He saw two processes as a clear fact of history. And it could be argued that Kuhn was led to important insights via his recognition of this distinction. Recently, Michael Friedman (2001) has argued that Kuhn was right and Quine was wrong on this point. If we approach actual scientific episodes using the idea of frameworks and the distinction between two kinds of change, we will be able to make more sense of how the sciences evolved. Quine's claim that we cannot make good sense out of a distinction between two kinds of conceptual change seems to be based on an overly stringent conception of what we would have to do in order to recognize two kinds of change.

Is Quine's denial of the two-process view justifiable? Thomas Ricketts, in a 1982 defense of Quine, tries to say in more detail how Quine might resist a two-process view. Suppose a scientist actually pins a set of basic principles on the laboratory door and insists that these principles constitute his framework. The principles on the door might change, the scientist says, but then they will change by a special process. Surely Quine will have to accept a two-process view in that case? Perhaps not. A scientist might pin some principles on the door and *say* they are different, but if Quine is right about the holistic nature of testing, the *actual* processes by which the scientist makes modifications to his beliefs will all be of the same kind. A pragmatic process of making adjustments to avoid tensions and unexpected observations is all that goes on, whether the ideas being changed are pinned on the door or not. There is no sense in which the moves made within the framework are "guided by facts" and those made to the framework are "merely pragmatic." All changes made to any belief result from the same kind of tinkering with the total network.

So in working out whether the two-process nature of scientific change is merely an illusion or not, we are led once again to fundamental questions about testing and confirmation. Still, I would have to say that despite these problems, the idea of a two-process theory of scientific change certainly *looks* like it has been useful to people working on specific cases. Alongside Kuhn we have Lakatos and Laudan, who have different kinds of two-process views. As I said earlier in this chapter, we can use the different theories of Kuhn and Laudan to make distinctions between different kinds of scientific fields—some fields are guided by Kuhnian paradigms, and some involve ongoing competition between research programs. Other fields might have a mixture of the two. These certainly *look* like useful distinctions. And Quine and his allies would concede that usefulness is the ultimate arbiter in a case like this.

This whole area has been the topic of lively discussion recently, and there is a lot more that could be said. The ghost of Immanuel Kant hangs over the whole discussion, as Kant was the first philosopher to develop a detailed view in which an abstract conceptual framework acts to guide empirical investigation ([1781] 1998). According to Kant, the basic framework that we bring to bear on the world is fixed and universal across all normal humans. We can never escape the framework (and would not want to if we could). Through the twentieth century, many philosophers found the idea of a "conceptual scheme" or framework appealing but insisted that these schemes can be changed and are not universal across cultures. Some of the radical ideas discussed in the next chapter can be seen as combining Kantian ideas about the role of conceptual schemes with a relativistic view in which alternative schemes are possible.

Not all philosophers of science can be neatly categorized as having "two-process" views or "one-process" views. Feyerabend is an interesting case. He recognized the psychological power of linguistic and cosmological frameworks, but he insisted that the imaginative person can *resist* the bounds of a framework. Popper, also, rejected the whole idea of frameworks as constraining thought and knowledge. He called it "the myth of the framework." Feyerabend did not see frameworks as mythical, but he thought that their bounds could be resisted and overcome.

Another interesting response to these issues is seen in Peter Galison's work (1997). Galison argues that what we often find in science is that fundamental changes in the different elements making up a scientific discipline are *not in step* with each other. Whereas Kuhn described a process in which there is simultaneous change in theoretical ideas, methods, standards, and observational data, Galison argues that within physics, fundamental change in experimental traditions tends to occur *nonsimultaneously* with fundamental change in theory. This is because of the partial autonomy of these different aspects of large-scale science. (Instrumentation is yet another tradition, with its own rates and causes of change.) So a big theoretical shift will be made more manageable by the fact that we can expect other aspects of the same field *not to be changing at the same time*. Disruptions happen more locally than they do on the Kuhnian model, and there are more resources available to the field to negotiate the transitions in an orderly way. The history of a scientific field shows "seams" of several different kinds, but these seams do not line up with each other. The structure as a whole is made stronger as a result.

Galison's picture shows us that there are many options for thinking about the relation between different kinds of scientific change. We should not think that there is a simple choice between the one-process camp (Popper, Quine, and Feyerabend in a sense) and the two-process camp (Carnap, Kuhn, Lakatos, Laudan, and Friedman). The situation is more complicated. And different kinds of frameworks have different roles—we should *not* think that Kant's universal conceptual framework has the same role as one of Laudan's research traditions! These are very different kinds of things.

I have also not said anything about the distinction between two-process views that see people as *modifying* their frameworks and those that see people as *jumping between* them. (Maybe in the end there is no difference.) In any case, the introduction and criticism of two-process views of conceptual change has been a recurring motif in the last hundred years of thinking about science and knowledge.

Further Reading

Lakatos's most famous work is his long paper in Lakatos and Musgrave, *Criticism and the Growth of Knowledge* (1970). Another key paper is Lakatos 1971. Cohen, Feyerabend, and Wartofsky 1976 is a collection on his work.

Feyerabend's most famous book is *Against Method* (1975), but his earlier papers are also interesting (collected in Feyerabend 1981). His later books are not as good, though *Science in a Free Society* (1978) has some interesting parts. *The Worst Enemy of Science* (Preston, Munévar, and Lamb 2000) is a collection of essays on Feyerabend. Horgan 1996 contains another great interview.

The Lakatos-Feyerabend relationship is documented in detail in Motterlini 1999.

Carnap's most famous paper on frameworks is his "Empiricism, Semantics, and Ontology" (1956), which is very readable, by Carnap standards. Another very influential (though difficult) discussion of "conceptual schemes," which follows up some of Quine's themes, is Davidson 1984.

8

The Challenge from Sociology of Science

8.1 Beyond Philosophy?

In the latter part of the twentieth century, ideas about science were in a state of flux. Options proliferated, including radical options. The previous chapter looked at some of these developments from the side of philosophy of science. The same phenomenon is found—perhaps even more so—in fields that border on philosophy. That is one topic of this chapter. I will focus on sociology of science, as this is an area that began to interact intensely with philosophy. Some of the same issues arose in the history of science, but it was sociology that sometimes set itself up as a replacement, or "successor discipline," to the philosophy of science (Bloor 1983).

8.2 Robert Merton and the "Old" Sociology of Science

Science is a social enterprise. It seems, then, that one field we should turn to in order to understand this fact is *sociology,* the general study of human social structures.

The "sociology of science" developed in the middle of the twentieth century. For a while it had little interaction with philosophy of science. The founder of the field, and the central figure for many years, was Robert Merton.

"Mertonian sociology of science" is basically mainstream sociology applied to the structure of science and to its historical development. In the 1940s Merton isolated what he called the "norms" of science—a set of basic values that govern scientific communities. These norms are *universalism, communism, disinterestedness,* and *organized skepticism.* Universalism is the idea that the personal attributes and social background of a person are irrelevant to the scientific value of the person's ideas. Communism involves the common ownership of scientific ideas and results. Anyone can make use of any scientific idea in his or her work; the French are not barred from using English results. The norm of disinterestedness is made questionable by Merton's later ideas, but the basic idea is that scientists are supposed to act for the benefit of a common scientific enterprise, rather than personal gain. Organized skepticism is a community-wide pattern of challenging and testing ideas instead of taking them on trust. (Merton sometimes added *humility* to his list of norms, but that one is less important.)

The four norms are one part of Merton's account of science. Merton added another big idea in a famous (and wonderfully readable) paper first presented in 1957. This is Merton's account of the *reward* system in science. Merton claimed that the basic currency for scientific reward is *recognition,* especially recognition for being the *first* person to come up with an idea. This, Merton claimed, is the only property right recognized in science. Once an idea is published, it becomes common scientific property, according to the norm of communism. In the best case, a scientist is rewarded by having the idea named after him, as we see in such cases as Darwinism, Planck's constant, and Boyle's law.

Merton argued that the importance of recognition is made evident by the fact that the history of science is crammed with priority disputes, often of the most acrimonious kind. And let no one suggest that the figures involved were always jealous also-rans. Galileo fought tooth and nail over recognition for various ideas of his. Newton fought Hooke over the inverse square law of gravity, and he fought Leibniz over the invention of calculus. The pattern has continued similarly from the seventeenth century to today. National loyalties are often a factor, as seen recently in the dispute between the American Robert Gallo and the Frenchman Luc Montagnier over the 1983 discovery of HIV. There are some exceptions to the general tendency; the most famous is the tremendously polite and gentlemanly non-dispute between Charles Darwin and Alfred Russell Wallace over the theory of evolution by natural selection in the nineteenth century. But the usual pattern when two scientists seem to hit on an idea around the same time is to fight for priority. As Merton says, the moral fervor seen in these debates, even on the part of those with no direct involvement, suggests that a basic community standard is operating.

Merton argued that the reward system of science mostly functions to encourage original thinking, which is a good thing. But the machine can also misfire, especially when the desire for reward overcomes everything else in a scientist's mind. The main "deviant" behaviors that result are fraud, plagiarism, and libel and slander. Of these, Merton held that fraud is very rare, plagiarism somewhat less rare, and libel and slander very common. Fraud is rare largely because of a rigorous internal policing by scientists, which derives partly from their own ambitions but also from organized skepticism. Plagiarism does happen, but the most usual outlet for deviant behavior is libel and slander of competitors. More precisely, what we find is a special form of slander that relates to the reward structure of science: *accusation* of plagiarism. This is vastly more common than actual theft. When two scientists seem to hit on an idea simultaneously, it is easy and often effective to insinuate that while *my* discovery was legitimate, it is no accident that Professor Z unveiled something very similar around the same time. After all, Professor Z was taking notes intensely during an informal talk I gave some months ago, and he cornered one of my graduate students and wanted to know how we had managed to . . . (etc.). Professor Z, or Z's allies, will often reply in kind.

Merton also has a poignant discussion of the fact that the kind of recognition that is the basic reward in science will only be given to a small number of scientists. There are not enough laws and constants for everyone to get one. The result is mild forms of deviancy such as the *mania to publish*. For pedestrian workers who cannot hope to produce a world-shaking discovery, publication becomes a substitute for real recognition.

Though the mania to publish is certainly real (and no less real in philosophy than in science), I suggest that Merton's analysis is not quite right on this point. Scientists (and philosophers) who cannot hope to produce another $e = mc^2$ will nonetheless often have real standing in a small community of people who work on the same detailed problems. Recognition even in a tiny community of colleagues can be a significant source of motivation. Kuhn's analysis of normal science recognized this fact. And in explaining the mania to publish, at least in recent years, university administrations and their desire for a simple way to measure productivity surely play a role.

Merton's editor Norman Storer suggests that we think of Merton's four norms as like a "motor" and think of the reward system as like the "electricity" that makes the motor run (Merton 1973). The norms describe a structure of social behavior, and the reward system is what motivates people to participate in these activities. But it is not so clear what the relation is between the two parts of the story. As Merton himself noted, the reward system can be at tension with the norms. In fact, I cannot see what remains of the norm of disinterestedness once we give Merton's analysis of the reward system. What we seem to have is not disinterestedness but a special *kind* of ambition and self-interest.

Earlier discussions in this book also suggest possible problems with Merton's "organized skepticism." There is definitely *something* right in this idea, and it has the same kind of intuitive appeal as simple statements of empiricism. But we must confront Kuhn's argument that too much willingness to revise basic beliefs makes for chaos in science. What we find in science, Kuhn claimed, is a delicate balance between skepticism and trust, between open-mindedness and dogmatism.

Still, we see in Merton's analysis a good *pattern* for a theory of the structure of science. We have a description of the rewards and incentives that motivate individual scientists, and we have an account of how these individual behaviors generate the higher-level social features of science. We will return to this idea in chapter 11.

Merton's sociology is often seen as the "old" style of sociology of science, a style that was superseded nearly thirty years ago. But there are some good ideas here, especially in the treatment of rewards. And sociology of science in the Merton tradition does continue, albeit with less drama and gnashing of teeth than we find in the newer approaches.

8.3 The Rise of the Strong Program

The sociology of science changed, expanded, and became more ambitious in the 1970s. Here is a standard way in which the "old" and "new" work are often distinguished. The older work wanted to describe the social structure and social placement of science as a *whole* but did not try to explain *particular scientific beliefs* in sociological terms. The newer approach has tried to use sociological methods to explain why scientists believe what they do, why they behave as they do, and how scientific thinking and practice change over time.

That standard description has some truth in it. But the newer sociology of science is also very interested in general norms, especially norms of reasonableness. Recent sociology of science has also worked with a different view of what scientific theories are like. At least in some of his work, Merton assumed a view of scientific theories that is close to logical empiricism—theories are basically networks of predictive generalizations. The newer sociology embraced Kuhn, holism about testing, incommensurability, new ideas about observation, and various speculative views about scientific language. In fact, these ideas make up a kind of "anti-positivist package" that was accepted not just by sociologists, but also by many historians, feminist critics, and others concerned with science in the latter part of the twentieth century.

The sociologists embraced some philosophers, but they intended their work to conflict with many traditional philosophical ideas about science. Some thought of sociology as replacing philosophy of science, a field that had become dried up and burdened with useless myths. The logical positivists, who seemed the most dried-up and abstract, became the bad guys, and eventually they turned into paradigm cases of Dead White Males.

The most famous project in this new form of sociology of science is the *strong program in the sociology of scientific knowledge*. This project was developed by an interdisciplinary group based in Edinburgh, Scotland, in the 1970s, headed (to some extent) by Barry Barnes and David Bloor. A central idea of the strong program is the "symmetry principle." This principle holds that all forms of belief and behavior should be approached using the *same kinds* of explanations. In particular, we should not give totally different kinds of explanations for beliefs that we think are true and beliefs that we think are false. Our own assessment of an idea should have no effect on how we explain its history and social role.

Applied to science, the symmetry principle tells us that scientific beliefs are products of the same general kinds of forces as other kinds of belief. Scientists are not some special breed of pure, disinterested thinkers who pay attention to nothing but data and logic. People of all kinds live in communities that have *socially established local norms* for regulating belief—norms for supporting claims, for handling disagreement, for working out who will be listened to and who will be ignored. These norms will often be subtle habits, rather than explicitly stated rules.

Scientists are people who work in an unusual kind of local community. This community is characterized by high prestige, lengthy training and initiation, notoriously bad fashion choices, and expensive toys. But according to the sociologists, it is still a community in which beliefs are established and defended via local norms that are human creations, maintained by social interaction. Scientists often look down on beliefs found in other communities, but this disparaging attitude is *part of* the local norms of the scientific community. It is one of the rules of the game.

As a result, we must recognize that the *kinds* of factors that explain why scientists came to believe that genes are made of DNA are the same *kinds* of factors used in explaining how other communities arrive at their very different beliefs—for example, a tribal community's belief that a drought was due to the ill will of a local deity. In both cases the beliefs are established and maintained in the community by the deployment of local norms of argument and justification. The norms *themselves* vary between the tribal community and the community of scientists, but the same general principles apply in both cases. Most importantly, we should not give the *Real World* a special role in the explanation of scientific belief that it does not have in the explanation of other beliefs that pass local community norms.

The strong program also sought to analyze particular scientific theories and their relation to social circumstances. This work became especially controversial. The aim was to explain some scientific beliefs in terms of the political "interests" of scientists and their place within society.

For example, Donald MacKenzie (1981) argued that the development of some of the most important ideas in modern statistics should be understood in terms of the role these tools had in nineteenth-century English thinking about human evolution and its social consequences. That connection went partly via the program of *eugenics,* the attempt to influence human evolution by encouraging some people to breed and discouraging others from doing so. MacKenzie argued that a body of biological, mathematical, and social ideas was well matched to the "interests" of the ambitious, reformist English middle classes. So he was asserting some kind of link between the popularity of specific scientific and mathematical ideas, on the one hand, and broader political factors on the other. What *kind* of link was this supposed to be? MacKenzie was cautious. When links are made between specific scientific ideas and their political context, the sociologist of science is quick to say that no simple *determination* of scientific thinking by political factors is being alleged. Sometimes metaphorical terms like "reflect" are used; scientific ideas will "reflect the interests" of a social group. Certainly it can sometimes be shown that the popularity of a scientific idea *benefits* a social group. But is this benefit supposed to *explain* the popularity of the scientific idea, or not? If so, is the explanation supposed to be a causal explanation, albeit a qualified one, or some other kind of explanation? This has been the source of some obscurity, but the issue of causal analysis in complex social systems is often very difficult. *Some* kind of explanation is intended, though.

This work on scientific ideas and "interests" antagonized conventional philosophers and historians, but it antagonized not only the "old farts." Even Kuhn was critical of it. Although Kuhn's work is always cited by those seeking to tie science to its broader political context, *Structure* did not have much to say about the influence of "external" political life on science. Kuhn analyzed the "internal" politics of science—who writes the textbooks, who determines which problems have high priority. But he saw an insulation of scientific decision making from broader political influences as a *strength* of science. Despite his status as a hero, Kuhn did not like the more radical sociology of science that followed him.

The strong program is also often associated with *relativism.* Many sociologists accepted this label, but we need to be careful. There are so many definitions of relativism floating around that the sense of relativism embraced by the sociologists need not be the same as that used by commentators and critics. The forms of relativism that are important here concern standards of rationality, evidence, and justification. Basically, relativism in this context holds that there is no single set of standards entitled to govern the justification of beliefs. The applicability of such standards depends on one's situation or point of view (see also glossary and section 6.3). In this sense, the strong program does tend to be relativist. It holds that science has no special authority that extends beyond all local norms. Instead, the norms and standards that govern scientific belief can be justified only *from the inside,* and that is true of other, nonscientific norms as well. We who live within science-dominated societies will find it compelling to say, "Science *really is* the best way of learning about the world." But, according to the strong program, saying those things is just an expression of our local norms. No one can hope to take a point of view *outside* all local norms and conceptual systems and say, This conceptual system or this set of local norms really is the best, the one that adapts us best to the world.

So, despite some differences within the field, it is fair to say that the strong program is an expression of a relativist position about belief and justification.

A famous problem for relativists is the application of relativism to itself. The problem does have various solutions, but it can definitely lead to tangles. Unfortunately, that is what happened in sociology of science. The application of the field's principles to itself led to interminable discussions that have weighed down the field. If all beliefs are to be explained in terms of the same kinds of social factors, and no set of local norms can be judged "really" superior from an external standpoint, then what about the theories found in sociology of science? This came to be called the "problem of reflexivity." Mostly the sociologists of science *accepted* that their claims were true of their own ideas. They accepted that their own theories were only justified according to local social norms. This conclusion is OK, but the whole issue led to endless methodological obsessing and navel-gazing.

In this section I have focused on a particular, dramatic strand in post-Merton sociology of science. But although it is easy to write as if "the strong program" was a clean and definite package, the program contained a good deal of variety. And it was not the only kind of sociology of science developing in this period. Just as the strong program elbowed aside earlier social accounts of science in the 1970s, it was to be partially elbowed aside in turn, in the 1980s.

8.4 Leviathan and Latour

This section will look at the two most famous works in recent sociology of science.

The first is a piece of sociologically informed history, rather than pure sociology: Steven Shapin and Simon Schaffer's *Leviathan and the Air Pump* (1985; I will abbreviate the book as *Leviathan*). This book does not advocate the strong program, but it is often seen as a sophisticated development of those ideas. The book is so widely respected, in fact, that various different camps tend to claim it as their own.

The second work is more controversial; it was important in a shift that took place in sociology of science: Bruno Latour and Stephen Woolgar's *Laboratory Life* (1979). This book appeared before *Leviathan;* it is famous as a pioneering work in its style.

Leviathan discusses the rise of experimental science in seventeenth-century England. This is seen as a pivotal case for our understanding of science, pivotal for its historical role in establishing the social structure that science has and also for illustrating this structure especially clearly. The book focuses on a dispute between Robert Boyle, a leader in the new experimental science, and Thomas Hobbes. Hobbes is remembered now mainly as a political philosopher (Hobbes's 1660 book *Leviathan,* which Shapin and Schaffer refer to in their title, is a work of that kind), but he also engaged in scientific disputes. The battle between Boyle and Hobbes was not "science versus religion" or anything like that; it was a battle over some specific scientific issues and over the proper *form* for scientific work and argumentation. Boyle prevailed.

According to Shapin and Schaffer, what came out of this period, and especially from Boyle's work, was a new way of bringing experience to bear on theoretical investigation. Boyle and his allies developed a new picture of *what* should be the subject of organized investigation and dispute, and *how* these disputes should be settled. The Royal Society of London, founded in 1660 by Boyle's group, became the institutional embodiment of the new approach. Boyle's approach did not become *the* model for science during the later seventeenth century, but it became one very important model, especially in England. There were some fairly strong differences in scientific "style" between different European countries during this period (and many would say that these have not entirely disappeared).

Boyle sought to sharply distinguish the public, cooperative investigation of experimental "matters of fact" from other kinds of work. Proposing causal hypotheses *about* experimental results is always speculative and should only be done cautiously. Theological and metaphysical issues should be kept entirely separate from experimental work.

In marking out a specific area in which dispute could be controlled and productive, Boyle hoped to show that scientific argument was compatible with social order. The seventeenth century had seen civil war in England, and this whole period in European history was one in which even the most abstract theological questions seemed capable of leading to violent unrest. So there was much concern with the problem of how to control dissent and dispute—how to stop it from spilling over into chaos. According to Shapin and Schaffer, Boyle saw his group of experimentally minded colleagues as a model for order and conflict resolution in society at large.

Boyle was not only setting up new ways of organizing work; he was also setting up new ways of *talking:* new ways of asking and answering questions, handling objections, and reaching agreement. We see this, Shapin and Schaffer argue, in Boyle's handling of key terms like "vacuum." The existence of vacuums was a key topic of debate in the seventeenth century. Aristotle's physics held that vacuums could not exist, but various lines of experiment suggested that perhaps they could. Boyle's experimental work involved the use of a pump that could apparently evacuate all or almost all the air from a glass container, in which experiments could then be performed. Shapin and Schaffer argue that Boyle was not really trying to answer the standard questions about vacuums. Instead, he was *reconstruing* questions about vacuums in a way that brought them into contact with his experimental apparatus. Critics could—and did—complain that Boyle's pump could not settle the questions they wanted to ask. Boyle's strategy was to subtly *replace* these questions with other questions that *could* be the topic of experimental work. The old questions—such as whether an absolutely pure vacuum could exist—had been set up in such a way that they would generate endless and uncontrollable dispute.

Shapin and Schaffer present their view in terms taken from the (later) philosophy of Ludwig Wittgenstein. Since Wittgenstein has influenced many people in the sociology of science, it is worth taking a moment to sketch the relevant ideas. Wittgenstein's early ideas about logic and language influenced logical positivism. His later ideas, especially his *Philosophical Investigations* (1953), were very different, and they had massive effect on late-twentieth-century thought. These later ideas are more an "anti-theory" than a theory; they are an attempt to show that philosophical problems arise from pathologies of language. Philosophy arises from a subtle transition between ordinary use of language and a kind of linguistic misfiring, in which questions that are really incoherent can seem to make sense. Wittgenstein wanted to diagnose and put an end to these misguided linguistic excursions. He avoided formulating *theories* of anything, but some of his ideas have been adapted for use in theories in various areas, including sociology of science (Bloor 1983).

Two ideas are especially popular. A "form of life" for Wittgenstein is something like a set of basic practices, behaviors, and values. Actions and decisions can make sense within a form of life, but a form of life as a whole cannot be justified externally. It's just the way a group of people live. Wittgenstein was not much interested in the kinds of cultural variation studied by sociologists and anthropologists, and it is not clear what sort of "unit" a form of life is for him. But sociologists have adapted the concept to fit the kinds of groups they study.

The second big concept drawn from Wittgenstein is the concept of a "language game." A language game is something like a pattern of linguistic habits that contribute to a form of life and make sense within it. Wittgenstein opposed a picture of language in which words and sentences are attached to their own particular meanings (mental images, perhaps) that determine how language is used. Instead, Wittgenstein claimed we should think of the socially maintained patterns of language use as *all there is* to the "meaning" of language. Shapin and Schaffer argue that Boyle's treatment of key terms like "vacuum" established a new language game. This language game was a key component in a new form of life, the form of life of experimental science.

At this point you may be remembering the logical positivists and their attempt to analyze the meaning of scientific language in terms of patterns in experience. Is the idea of a language game developed to serve experimental science different from this positivist idea? It is different. The logical positivists claimed that the right theory of meaning would show that all meaningful language *ever* does is describe patterns in experience. According to Shapin and Schaffer, Boyle was setting up a *new* way of using language. So there is perhaps more of a connection to the "operationalism" of the physicist Bridgman, who was briefly mentioned in chapter 2. Bridgman (1927) urged that scientists *reform* their use of language to ensure that each term has a direct connection to empirical testing.

Those are the central ideas in *Leviathan and the Air Pump*. But something should be said about another, more problematic, feature of the book.

Shapin and Schaffer claim that Boyle and other scientists are engaged in the *manufacture of facts*. In everyday talk the phrase "manufacture of facts" would be taken to indicate deception, but that is not what Shapin and Schaffer have in mind. For them, there is nothing bad about the manufacture of facts; they want us to get used to the idea that facts in general are *made rather than found*. This is reminiscent of Kuhn's claim, in Chapter X of *Structure*, that the world changes during a scientific revolution. Like many others who use these terms, Shapin and Schaffer want to reject a picture of the scientist as a passive receiver of information from the world. But a denial of passivity does not require this kind of talk, and it often leads to trouble. For example, at the very end of *Leviathan*, their discussion of "making" leads Shapin and Schaffer to express their overall conclusions in a way that involves a real confusion. They say: "It is ourselves and not reality that is responsible for what we know" (1985, 344). This is a classic example of a false dichotomy. Neither we alone nor reality alone is "responsible" for human knowledge. The rough answer is that *both* are responsible for it; knowledge involves an interaction between the two. Even this formulation is imperfect; human knowledge is *part* of reality, not something separate from or outside it. But, speaking roughly, in order to understand knowledge, we need both a theory of human thought, language, and social interaction, *and* a theory of how these human capacities are connected to the world outside us.

I now move on to a second famous work in the sociology of science, Latour and Woolgar's *Laboratory Life* (1979). In the mid-1970s, Bruno Latour, a French sociologist, spent a couple of years visiting a molecular biology laboratory, the Salk Institute in San Diego. He went as a charming observer who knew little about molecular biology. During the time that Latour was there, the lab did work that resulted in a Nobel Prize; they discovered the chemical structure of a hormone involved in the regulation of human growth. Latour wrote *Laboratory Life,* with Steven Woolgar, as a description of the lab's work.

Latour and Woolgar, in their account, *ignored* most of what a normal description of a piece of science would focus on. They ignored the state of our knowledge of hormones; they ignored the ways in which experimental methods in the field are able to discriminate alternative chemical structures; they ignored how the new discovery fitted into the rest of biology. Instead, Latour looked at the lab in a sort of deliberately superficial and self-contained way. The lab was a kind of machine where chemicals, small animals, and reams of blank paper came in at one end, and small printed pieces of paper—journal articles and technical reports—came out at the other end. In between the two, a huge amount of "processing" went on, processing that turned the mass of raw materials into the intricate finished products (see fig. 8.1).

Latour saw this processing as aimed at taking scientific claims and building structures of "support" around them, so they would eventually be taken as *facts.* A key step in this process is *hiding* the human work involved in turning something into a fact; to turn something into a fact is to make it look like it is not a human product but is given directly by nature.

Laboratory Life was a huge success. To many it seemed like a breath of fresh air, a book that exuded wit and imagination. Along with other works, it prompted a shift within the field. The strong program came to seem crude. The strong program wanted to get rid of explanations of scientific belief in which nature just stamps itself on the minds of the scientific community. But perhaps the strong program was replacing this with an equally crude picture, in which social and political "interests" stamp themselves on the scientific community. This is not a very fair reading of the strong

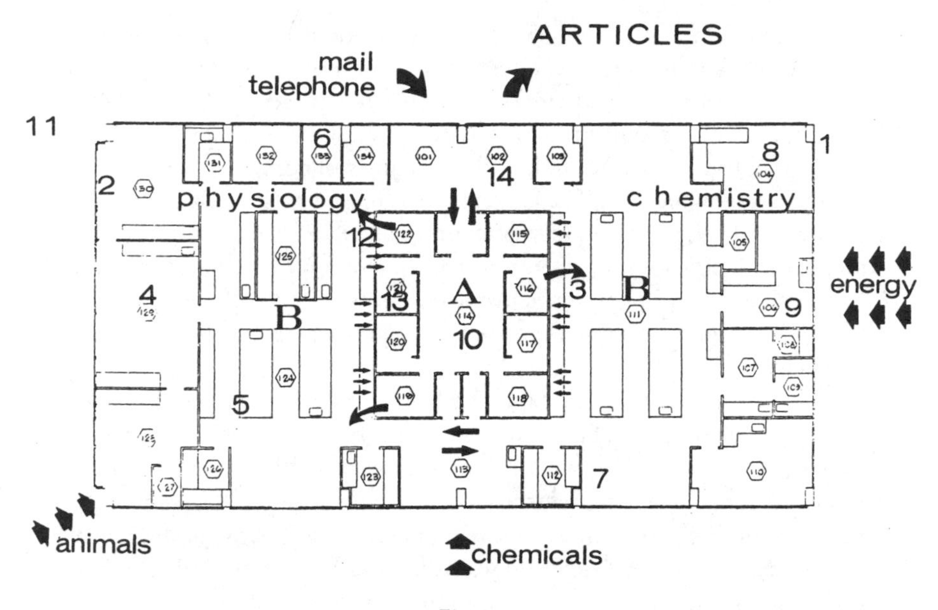

Fig. 8.1
The lab according to Latour and Woolgar (From *Laboratory Life: The Construction of Scientific Facts,* by Bruno Latour and Steve Woolgar, © 1986; reprinted by permission of Princeton University Press)

program. The sociologists were being caricatured, just as they had caricatured traditional philosophy of science! Some would see justice there. Latour also inspired a different *style* in sociology of science, a style that might be described as elusive, self-conscious, and literary.

Latour's approach, which borrows from French philosophy, sociology, and semiotics, is sometimes called the "actor-network" theory. What the sociologist does is to study the fine structure of the internal dynamics of scientific work, especially dispute and negotiation about what has been established. The sociologist does not begin the story by taking for granted "pressures" or "interests" in society at large, and "nature" or the real world is not taken for granted either. Instead, both "society" and "nature" are seen as *products,* not *causes,* of the settlement of scientific controversies (Latour 1987). Where traditional empiricist philosophy had seen science as "data-driven," and the strong program had seen science as "interest-driven," Latour sees scientific work *itself* as the driver.

In Latour's view, when we explain why one side succeeded and another failed in a scientific controversy, we should never give the explanation in terms of nature itself. Both sides will be claiming that they are the ones in tune with the facts. But when one side wins, that side's version of "the facts" becomes *immune to challenge.* Latour describes this final step as a process in which facts are *created,* or *constructed,* by scientific work.

There is an interesting strategy in Latour's work here, along with much obscurity. Latour gets us to look at the dynamics of controversy in science in a very fine-grained way. What social role is played by appeals to "truth," "nature," and "the facts"? How do people *use* these terms before, during, and after the settlement of disputes? These are good questions. One kind of understanding we might have of the concepts of "truth" and "nature" is an understanding of how they are used as *resources* in arguments and discussion.

An investigation of this kind might tell us a lot about how people decide what they *take* to be real, but this does not mean that the settlement of scientific disputes determines what *is* real. However, Latour refuses to make this distinction when he presents his work. To some extent this seems to be due to his interest in very unorthodox philosophical positions. But sometimes his neglect of this distinction seems to be just a matter of his dashing, provocative writing style (a style common in French intellectual life).

Revolutions, as is well known, have a habit of eating their children. Although Latour is often seen as displacing or digesting the strong program, that program has been showing some resistance to the French stomach acids. A striking attack on Latour was recently published by the strong programmer David Bloor (1999). Bloor's paper radiates exasperation at the fact that Latour's obscure project has come to seem the more sophisticated and appealing option. Bloor urges a return to the strong program, and he presents that program in a way that avoids careless talk about the "construction of reality." Bloor is an exception; sociology of science has rarely treated this issue with care. And although I doubt that the strong program is the way of the future, Bloor is right that recognizing the role of social structure in science does not require strange inversions of the relations between thought and reality.

In both its radical work and its more cautious work, sociology of science in the latter part of the twentieth century tended to suggest an unusual *picture* of science. This is a picture in which science is controlled entirely by human collective choices and social interests. What makes science run is negotiation, conflict resolution, hierarchies, power inequalities. . . . There seems to be no place in the picture for the responsiveness of scientific belief to the real structure of the world being investigated. Often sociologists accept that *of course* the real world imposes *some* constraints on what we believe. But any particular observations that a person might make are always subject to so much reinterpretation, reconstruing, filtering, and negotiation that they cannot *guide* belief or theory change. What makes things happen in science—what makes people believe one theory rather than another—is the interaction of social forces.

Further Reading

A standard collection of Merton's works is *The Sociology of Science* (1973). A central work in the rise of the strong program is Bloor's *Knowledge and Social Imagery* (1976). See also Barnes, Bloor, and Henry 1996. Shapin 1982 is a good survey of historical work done by sociologists of science. On the issue of relativism, see Barnes and Bloor 1982 and other papers in that collection (Hollis and Lukes 1982).

Shapin has followed up the argument in *Leviathan* with some other very interesting work, especially *A Social History of Truth* (1994). Latour's other famous books include *Science in Action* (1987), *The Pasteurization of France* (1988), and *We Have Never Been Modern* (1993).

9

Feminism and Science Studies

9.1 "Science Is Political"

The relationship between science and politics was subjected to new kinds of scrutiny in the late twentieth century. To some extent, the overall *image* of science changed, especially in the humanities. Generalizations are risky here, but we might say that through much of the last three hundred years, science has been regarded in Western societies as a progressive, anti-authoritarian force, able to challenge and break down entrenched ideas and arrangements. This view was most vivid in the "Enlightenment" period, in the eighteenth century; confidence in science as a progressive force is one aspect of what are now called "Enlightenment values." There have always been exceptions to this cultural image of science, within nineteenth-century romanticism, for example, and in some Marxist thought. But various parts of intellectual culture saw a larger shift in attitudes toward science in the latter part of the twentieth century. The cold war was a crucial cause of this shift, but there were other currents at work also.

Science came to be seen instead as a force in the maintenance of the "status quo," especially with respect to political inequalities. On the side of politics that considers itself progressive rather than conservative, many began to treat science as part of a larger, multi-tentacled political structure that acts to reinforce subtle forms of exclusion and coercion, even in apparently "free and democratic" societies. The anti-authoritarian image of science came to be seen largely as just "good PR." And the institution of science itself, it was argued, is full of hidden features that include some individuals and exclude others.

Many thought that by showing the connections between scientific institutions and political power, it would become clear that "science is political," rather than being an institution outside of politics that enjoys a special authority derived from this political neutrality. Revealing the political embedding of science would also have relevance to questions about education, medicine, and a variety of other crucial areas of social policy.

The most important manifestation of this new attitude is found in the development of feminist critiques of science and feminist philosophies of science. That is the topic of the first part of this chapter. The second part looks at the growth of another new approach to understanding science, the interdisciplinary field known as "Science Studies."

9.2 The Man of Reason

Feminist thinking about science makes up a diverse movement. It is unified, perhaps, by the idea that science has been part of a structure that has perpetuated inequalities between men and women. Science itself, and mainstream theorizing about science and knowledge, have helped to keep women in a "second-class" position as thinkers, knowers, and intellectual citizens. (Even these generalizations about feminist discussions of science have exceptions.) According to feminist analyses, society has suffered from this, and so has science itself. So reform of some kind is needed. There is disagreement on the appropriate kind of reform—ranging from simple suggestions like the inclusion of more women in the sciences, through the encouragement of a specific kind of female "voice" in science, to dethroning science from its preeminent position in Western culture. Feminist thinking about science was often allied with work in the sociology of science, and Kuhn, Feyerabend, and Wittgenstein were also seen as helpful. Some feminists made more unfortunate alliances with Freudian psychoanalysis.

We should distinguish feminist *philosophical* ideas about science from more basic feminist *political* ideas. Feminism in general aims to understand and fight against inequalities between the sexes, with respect to political rights, economic standing, and social status. This has a simple application to science: women were for many years excluded or discouraged from a life in science, as they were excluded from other high-prestige areas of work. This is a simple matter of equality of opportunity, one that raises questions about policy (such as the appropriateness of affirmative action) but does not raise issues in the philosophy of science itself.

Other feminist work did engage with philosophical issues about science. The work might be categorized in terms of three overlapping strands. One strand is feminist analysis in the history of ideas and the history of science. A second is feminist analysis of specific scientific fields and theories, especially in social science, biology, and medicine. The third is *feminist epistemology,* the attempt to analyze rationality, knowledge, and other basic epistemological concepts from a feminist point of view. Here I include analysis of the social structure of science, when that work bears on epistemology.

I will start by discussing a book written fairly early in the tradition, Genevieve Lloyd's *The Man of Reason* (1984). Lloyd analyzes the historical roots of ideas about knowledge and rationality and also draws conclusions for epistemology. The discussion that is specifically relevant to us is found in the early chapters, where she considers figures such as Plato, Aristotle, Descartes, and Bacon. The book illustrates what I think has been a common pattern. Lloyd tells a very interesting—sometimes compelling—story in the history of ideas. But it is harder to work out what consequences these historical facts have for epistemology.

Lloyd argues that the early development of ideas about reason and knowledge were greatly affected by views about the relation between maleness and femaleness. The concept of reason evolved in Western philosophy in a way that associated reasonableness with maleness, and associated the female mind with a set of psychological traits that *contrast* with reasonableness.

A key source for this pattern of thinking, according to Lloyd, is the old association between femaleness and nature; the earth is fertile, female, the source of life. Via this association, ideas about the relationship between the mind and nature were modeled on the relationship between male and female. The relations between the sexes also provided a model for theorizing about the relations between different aspects of the mind itself—between perception and thought, and between reason and emotion. The upshot was that the ideas feeding into the early development of science and philosophy in Europe incorporated, in various different forms, an association between reason and maleness. And the development of the idea of femininity was shaped by an *opposition* between femininity and reason. Femininity was associated with receptivity, intuition, empathy, and emotion.

Lloyd's best example is the case of Francis Bacon, the seventeenth-century English thinker who wrote extensively about the new empirical methods of investigation and their promise for mankind. Bacon attacked the ancient Greek picture of knowledge as contemplation. For Bacon, real knowledge is manifested in *control* of nature: "Knowledge is power." But as Bacon developed this idea, he retained the image of nature as female. His model for the relation between the mind and nature was the model of *marriage,* a marriage between the knower (man) and nature (woman). The features of a good marriage, as run by the man, correspond to the features of successful knowledge of the world. So what is a good husband like? A good husband is respectful, but he is also firm and definitely in charge. The scientist approaching nature should approach her with respect and restraint. But control is certainly needed; "Nature betrays her secrets more fully when in the grip and under the pressure of art than when in enjoyment of her natural liberty." And the products of what occurs on the "nuptual couch" will be useful knowledge for the improvement of mankind (quoted in Lloyd 1984, 11–12. Some other feminists have been far tougher on Bacon than Lloyd was: see Harding 1986).

Cases like this suggest that views about the relations between men and women were important resources in the development of ideas about reason and knowledge. Although the question is difficult, it surely seems likely that these associations did affect both the lives of women and the path taken by science in the early modern period. The harder question is what philosophical consequences these historical facts have for us now, given the massive changes to political life and to science since then. It is not hard to find a residue of these old associations embedded in metaphors that are still around. To pick a simple case, scientists constantly talk about whether or not a phenomenon will "yield" to a particular method of analysis. To my ears (though not to everyone's), this metaphor always has a resonance of sexual conquest. But whether these metaphors have much *effect* on either society or science today is a more difficult issue.

Evelyn Fox Keller is one feminist who thinks there is a real problem here. She holds that the general picture of science we have inherited has real effects on women entering science; the woman scientist has to choose between "inauthenticity" and "subversion." The concept of authenticity is a subtle one drawn from existentialist philosophy, but Keller illustrates her point with an analogy: "Just as surely as inauthenticity is the cost a woman suffers by joining men in misogynist jokes, so it is, equally, the cost suffered by a woman who identifies with an image of the scientist modeled on the patriarchal husband" (2002, 134–35).

9.3 The Case of Primatology

I turn now to a case that many see as a good, clear example of the role that gender has played in a particular part of science. More specifically, this is often seen as a case in which the gender of researchers has had an effect on the development of ideas, one where science has benefited from an increasing role for women in the field. The example concerns the last thirty years or so in the study of social behavior, especially sexual behavior, in nonhuman primates like chimps and baboons. These phenomena are studied (with slightly different emphases) in the fields of primatology and behavioral ecology.

These parts of biology initially developed a picture of primate sexual life in which females were seen as rather passive. Social life, and sexual life in particular, were regarded as controlled, sometimes cruelly, by males. That picture was linked to some important pieces of "high theory" in evolutionary biology. In many animals, although by no means all, there is a great deal of variation across individuals in male reproductive success, and less variation in female reproductive success. This is a consequence of the fact that one male can, in principle, impregnate large numbers of females. As it is often said, "sperm are cheap." Female reproductive success is limited in many animals by the high costs of pregnancy.

This kind of asymmetry between the sexes is of considerable evolutionary importance in the organisms in which it is found. But it has often been used in rather simplistic patterns of explanation, without regard for many ways in which its effects can be modified by other factors. In early primatology, it was taken to support a view holding that male sexual behavior had been finely honed by natural selection, while female behavior had not, because females could do much less to affect their reproductive success.

According to Sarah Blaffer Hrdy (2002), this picture began to shift in the 1970s. Careful observation revealed a far more active and complex role for female primates. It became apparent that many female primates have elaborate sex lives, involving a lot more different kinds of sexual contact than one would expect, based on the old picture. Females seem to engage in subtle patterns of manipulation of male behavior, and much of the manipulation may be directed at influencing male behavior toward offspring. The basic theoretical idea that the high potential variance in male mating success has large effects on the evolution of behavior still stands, but there is now a much more sophisticated picture of the interaction between this factor and other factors, especially the strategies available to females.

This shift in thinking within primatology coincided, at least roughly, with an influx of women into the field. Primatology is, in fact, one of the scientific fields in which the presence of women is unusually strong. What role did the presence of women have in changing opinions within the field? According to Hrdy (and according to others I have spoken to), the idea that this increasing representation of women had a significant role in shifting people's views about female primate behavior is fairly routinely accepted within primatology. Hrdy adds that this view seems to be accepted more in the United States than in Britain (2002, 187). Hrdy herself is rather cautious about this issue, but she suggests that women researchers, like herself, did tend to empathize with female primates and watched the details of their behavior more closely than their male colleagues had.

9.4 Feminist Epistemology

Let us now look more closely at feminist epistemology, or rather, the part of feminist epistemology that deals with science. This is a diverse and sometimes difficult field. It includes work that uses feminist theory as a basis for criticizing how science handles evidence and assesses theories. It also includes feminist criticism of the social structure and organization of science, where that structure affects epistemological issues. Most ambitiously, some feminist epistemologists have argued that even our fundamental concepts of reason, evidence, and truth are covertly sexist. Feminist epistemology also goes beyond criticism to make suggestions about reform—how to make science better at finding out about the world (if that goal is to be retained), and also how to make science more socially responsible.

In discussing some of the options here, I will modify some categorizations used by Sandra Harding (1986, 1996). Harding distinguishes three kinds of feminist criticism of science. The earliest and least controversial she calls *spontaneous feminist empiricism.* This is the project of using a feminist point of view to criticize biases and other problems in scientific work, but doing so in a way that does not challenge the traditional ideals, methods, and norms of science.

Harding's second category is *philosophical feminist empiricism.* Helen Longino's work (1990) is probably the most influential within this camp, and I will discuss it below. Here the aim is to revise and improve traditional ideas about science and knowledge, but to do so in a way that remains faithful to the most basic empiricist themes. Relativism is to be avoided. One hope is that more sophisticated criticisms of particular scientific practices will result.

The third category I will call *radical feminist epistemology.* Two main approaches might be distinguished within this grouping. One is what Harding calls *feminist postmodernism.* This work tends to embrace relativism. Members of different genders, different ethnic groups, and different socioeconomic classes see the world fundamentally differently. The idea of a single "true" description of the world that transcends these different perspectives is a harmful illusion.

The second radical approach is *standpoint epistemology.* This is not a relativist view; it is more ambitious than that. Standpoint epistemology stresses the role of the "situatedness" of an investigator or knower—their physical nature, location, and status in the world. The idea is that while traditional epistemology has seen "situatedness" as a potential problem for an investigator, in fact it can be a strength. Standpoint theory holds that there are some facts that are only visible from a special point of view, the point of view of people who have been oppressed or "marginalized" by society. Those at the margins, or the bottom of the heap, will be able to *criticize the basics*—both in scientific fields and in political discussion—in a way that others cannot. Science will benefit from taking more seriously the ideas developed by people with this special point of view. This is not a relativist position because the marginalized are seen as *really* having better access to crucial facts than other people have.

One of the main debates in feminist epistemology has been between forms of philosophical feminist empiricism and views that are more radical, especially standpoint epistemology. The arguments for more radical options have not been convincing. Standpoint theory holds that the experiences of marginalized people have special value. If that is right, what *sort* of value is this? As Longino argues, it is not likely to be a *general* superiority of a kind that would justify our treating a marginalized point of view as *the* most important or reliable. If some facts are more visible to the marginalized and oppressed, other facts will surely be more visible to the privileged. The experiences of the marginalized are more likely to be valuable *as a special kind of input* into discussion and argument. So the right way to think here is in terms of a "pool" of different ideas, contributed by those with different points of view. Longino argues that the picture that results is a revised version of empiricism.

Longino calls this revised view "contextual empiricism." This is a form of empiricism that emphasizes the role of social interaction. Longino argues that in order to be able to distinguish rationality from irrationality we should take the *social group* as our basic unit. Science is rational to the extent that it chooses theories from a diverse pool of options reflecting different points of view, and makes its choice via a critical dialogue that reaches consensus without coercion. Diversity in the ideas in the pool is facilitated by diversity in the backgrounds of those participating in the discussion. Epistemology becomes a field that tries to distinguish good community-level procedures from bad ones.

If this is the right way to incorporate feminist ideas into epistemology, it is a way that follows a fairly old tradition (as Longino would not deny). Paul Feyerabend, as we saw in chapter 7, argued for the importance of maintaining diversity in scientific communities. And as Elisabeth Lloyd argues, Feyerabend was extending and radicalizing a line of argument from John Stuart Mill (Lloyd 1997). Diversity, for Mill, provides the raw materials for social and intellectual progress, via a vigorous "marketplace of ideas."

The idea that a diversity of viewpoints improves critical discussion is definitely appealing. The role of gender in the mix is a separate question, as writers like Longino accept. Is it really true that men and women in modern Western societies have different perspectives of a kind that is relevant to science? Feminists accept that other differences, especially class differences and ethnic differences, may have as much of an effect as do gender differences, or even more than that. But many feminists expect there to be some definite "patterning" in the great soup of intellectual diversity that is due to gender differences.

So might we expect women to have a different *style* of theorizing or reasoning that derives from their different experience? Certainly there will be some facts that women will tend to have a different perspective on. The physical experience of being a woman or a man will make a difference to how some aspects of life are experienced. And at least for the near future, the early education and acculturation of girls and boys will have this effect as well. But we should be careful of claims that go far beyond this. It is a much harder question whether or not the experience and viewpoint of women is systematically different from that of men *in a way that is likely to matter to scientific disputes*. There is a risk of lapsing into simplistic generalizations here.

This problem takes us back to some issues discussed in the previous section. In some situations it can be argued that a particular bias, or neglect of options, found in a scientific field may be due to gender. Primatology is one area where this argument has been taken seriously. It is also possible to go beyond this and argue that there is a distinctive way of thinking, and interacting with the world, found in women scientists. If so, this might be partly due to a distinctive way that women tend to think, and it might also be partly due to their situation and experience in male-dominated fields. A famous example is Evelyn Fox Keller's work on Barbara McClintock, the geneticist who discovered "jumping genes" that move around within the genome of an organism. The jumping-genes idea was for some time considered to be a very strange hypothesis, but McClintock turned out to be right. McClintock was very much an outsider in genetics, and Keller also argues that McClintock had a "feeling for the organism" that enabled her to do a different style of science from that of her male colleagues (1983). Keller is rather cautious in her claims here; she does *not* want to argue that there will be a "sharp differentiation" between women's and men's work in science (2002, 134). But she does seem to think there will be some systematic differences. Many would object to the suggestion that a "feeling for the organism" is likely to be an example, however. A case could be made that this psychological trait is found in many good biologists and that it has nothing to do with gender. Feminists themselves (including Keller) are also very wary about the possibility of contributing to a stereotyping of female contributions to scientific thinking. ("We must have a woman on this team, Jim, so someone will pick up on the holistic, interconnected stuff that might be going on in these reactions!")

Here I have discussed possible differences in "theoretical style" between men and women. Another possibility is that women will tend to bring a different kind of social interaction to scientific communities. Feminists have sometimes suggested that women are, on the whole, less competitive and more cooperative than men, though many feminists would now want to avoid simple generalizations of this kind (Miner and Longino 1987). If there are any differences of this nature, they may have important consequences for science. The next chapter will discuss the relation between cooperation and competition within science in detail, so we will return to the issue of gender differences then.

9.5 Science Studies, the Science Wars, and the Sokal Hoax

One of the main themes in this chapter and the previous one has been the constant expansion of the range of fields seeking to contribute to a general understanding of science. The two examples I have discussed in detail are sociology of science (chapter 8) and feminist criticism (this chapter). As well as this expansion, there has been a blurring of disciplinary boundaries. During the 1980s a number of workers decided to embrace this trend and create a new approach to studying science that would draw on many different fields without worrying about "whose questions were whose."

The resulting field is generally known as "Science Studies." The mixture has come to include not only history, sociology, and philosophy but also cultural anthropology, classics, economics, some parts of literary theory, feminist theory, and more marginal fields like semiotics, cultural studies, and critical theory. The aim is to draw on pretty much any field that can contribute our understanding of how science developed, how it works, and what role it has. Recently, the study of *technology*, as distinct from science, has sometimes been explicitly added as a goal.

The result of this reorganization has not been a massive breakthrough (as some might have hoped), and it has not been a disaster either (I will explain below why it might be seen that way). The history of recent thinking about science does show that there are good opportunities for cross-fertilization, borrowing, and joint work in this area. But it is not likely that the boundaries between fields will really disappear; philosophers, historians, sociologists, and literary theorists do look at the world somewhat differently. So, of course, we find a mixture of styles of work within Science Studies, ranging from the most sober, intricate historical research to wild flights of fancy that make Bruno Latour look like Rudolf Carnap. I am not denying that there are some distinctive tendencies and emphases in the field as a whole, however; one will be discussed at the end of this chapter.

Some of the most controversial work in Science Studies is allied to the notorious movement in the humanities known as "postmodernism" (Harvey 1989, Lyotard 1984). Postmodernism is a family of ideas and projects, ranging from architecture through art, history, and philosophy of language. The themes that are relevant to us here have to do with representation and meaning. Postmodernism is part of a recent tradition in the humanities that opposes the idea that language should be analyzed as a system used to *represent*, or "stand for," objects and situations in the world. This antirepresentationalist view of language influenced a lot of literary theory, as well as other humanistic disciplines, in the latter part of the twentieth century. Postmodernism is a spectacular outgrowth of that line of thought.

Sometimes postmodernists seem to be arguing that we, right now, live in a special time in history. We live in a time when a representational role for symbols is being replaced by a new role. The sea of symbols and languages in which we live, and their role in politics and in consumer culture, has *undermined* ordinary representational relations between symbols and objects. In understanding the role of symbols in our lives, it is no longer useful to apply concepts like accuracy, reference, and truth: behind every symbol lies not a real object, but another symbol. At other times postmodernism seems to become a tremendously obscure way of arguing for extreme forms of relativism, sometimes for a kind of skepticism and do-nothingism, and for extravagant metaphysical views about how language and reality are related.

Science Studies was rather welcoming to postmodernism and other adventurous ideas from the humanities. That did *not* mean that the sober historians of science stopped doing their sober, rigorous work; Science Studies is a diverse entity. However, the relationship between the new approach to science and obscure trends in the humanities affected the *image* that Science Studies came to have. And in time, there came a backlash.

The backlash occurred in the form of an attack both on Science Studies and on recent work in the humanities more generally. Some of the backlash arose within science itself; scientists were alarmed at the picture of science being presented to the broader culture. But much of the heat and noise was due to commentators' criticizing larger tends within academia and education. The perception was that science itself was under threat.

The resulting clash became known as the "Science Wars." Science Studies, and other work covered in this chapter, became a key battle ground. Some of the attacks on this work came from the side of conservatism in political and social thought. Advocates of "traditional" education, both in schools and in universities, worried that transmission of the treasures and values of Western civilization was being undermined by radical leftist faculty members in universities and soft-minded administrators in schools. The humanities had gone to hell, and now they were trying to wreck science as well, via endless relativist bleating that science is "just another approach to knowledge with no special status."

Although some of these battles had a simple political structure, the most influential and interesting episode did not. In 1994 an American physicist, Alan Sokal, submitted a paper to a literary-political journal called *Social Text,* which was doing a special issue on science. The paper was a parody of radical work in Science Studies; it used the jargon of postmodernism to discuss progressive political possibilities implicit in recent mathematical physics. The title of the paper gives a sense of the style: "Transgressing the Boundaries: Toward a Transformative Hermeneutics of Quantum Gravity." The argument of the paper was completely ridiculous and often quite funny. The aim was to see if the paper would be accepted and printed by the journal; Sokal believed that this would show the field had lost all intellectual standards and would print anything that used the right buzzwords and expressed the appropriate political sentiments.

Social Text published the paper (Sokal 1996b), and Sokal revealed his hoax in the journal *Lingua Franca* (1996a), an irreverent journal of academic life (sadly, defunct, at least for now). The uproar reverberated across the academic world and also made the newspapers. One of the things that made Sokal's attack so effective was that he was *not* writing from the point of view of conservative politics. He presented himself as a left-winger who felt that the Left had lost its way. The siren song of trendy French philosophy and literary theory had led the Left, and "progressive" politics more generally, away from its earlier alliance with science and landed it in a useless and pretentious quagmire.

Many philosophers in the English-speaking world felt vindicated by the Sokal hoax. Although English-speaking philosophy had produced radical ideas about science, for the most part it had not accepted postmodernism and other French-influenced literary-philosophical movements. Jacques Derrida, perhaps the most famous figure in all the humanities during this period, had never been embraced by the philosophical establishment and was regarded by many as a virtual charlatan. Philosophers thought their own journals were "hoax-proof" because of the philosophical demand for clear argumentation. (I do not know whether this conviction has been tested.)

Some mainstream philosophers of science, who had been made to look dried-up and boring by decades of racier work in neighboring fields, were elated. At the 1996 meetings of the Philosophy of Science Association, the presidential address was given by Abner Shimony, a senior philosopher of physics. Shimony's address was a reassertion of Enlightenment values, the values of science, democracy, rationality, equality, and secularism. Shimony called Sokal "a hero of the enlightenment" for his work in unmasking the foolishness of radical Science Studies.

Although some philosophers felt vindicated, others felt that damage had been done. In the discussion period after Shimony's talk, Arthur Fine and Philip Kitcher, two other prominent philosophers of science, lamented that after years spent bridging gaps between disciplines and establishing dialogue, Sokal's work was likely to polarize everything again. This fear was quite reasonable, as there had often been distrust between some fields. Philosophers might cease to pay any attention to work in neighboring fields, in the belief that they had lost all intellectual standards. Sociologists, on the other side, were likely to think that the underlying conservatism of philosophy had been revealed again; after all, they would say, the smug philosophers had sided with Sokal's cheap shot.

Science Studies was not seriously damaged by the Sokal hoax, but there have been some lasting effects. As I stressed earlier, the field was always diverse, even though its image to outsiders was sometimes dominated by the most high-risk work. There is less tolerance now for very jargon-laden and obscure writing. This is a good thing, and it is reason enough to be glad of what Sokal did. Internal obsessing about how Science Studies should be conducted is excessive, but it was excessive well before Sokal. More importantly, the fear that the gaps between different fields would widen dramatically has not been realized.

I have emphasized the mix within Science Studies of "straight" history with the most "bent" literary analysis of science and culture. But the field does exhibit some general tendencies. One is especially relevant here. Science Studies is rather hostile toward the idea of explaining patterns in scientific change in terms of relations between scientific theories and the structure of the world. Kuhn and the sociological work discussed in chapter 8 have left an enduring mark here. The explanations that are most emphatically rejected by Science Studies are explanations of the popularity of a theory in terms of its real accuracy or explanatory power. Explaining the *historical* role of a theory in terms of our *present* estimation of its worth is taken to be a bad mistake. And more generally, Science Studies is suspicious of the whole idea of looking at scientific theories in terms of how they relate to the preexisting structure of the world itself. What results is a gap in the account of science that Science Studies provides. After we have described the social structure of science itself, we need to also understand how that social structure and its products connect to the larger natural world within which scientific activity is embedded. This will be one of the themes in the chapters to follow.

Further Reading

Keller and Longino, *Feminism and Science* (1996), and Janet Kourany, *The Gender of Science* (2002), are both useful collections. The latter is quite comprehensive and includes the Hrdy paper I use in section 9.3. Hrdy's book *The Woman That Never Evolved* (1999) is a more detailed discussion of her ideas. Donna Haraway's *Primate Visions* (1989) is a very detailed historical and sociological discussion of primatology from a feminist point of view. For another interesting feminist case study, see Elisabeth Lloyd's work on theories of the evolution of female orgasm (1993).

Harding, *The Science Question in Feminism* (1986), and Longino, *Science as Social Knowledge* (1990), are two of the most influential books in feminist epistemology as applied to science. The *Monist* had a special issue on feminist epistemology in 1994.

Mario Biagioli, *The Science Studies Reader* (1999), is a good collection that illustrates the diversity of work in that field. For the Science Wars, see Gross and Levitt, *Higher Superstition* (1994), which includes criticisms of Bloor, Latour, Shapin, Schaffer, Harding, Longino, and various others I have discussed in these chapters. See also Koertge, *A House Built on Sand* (1998). The Sokal hoax is the subject of a book (of that name) edited by the *Lingua Franca* editors (2000). There is also a mass of material about the Sokal hoax on the World Wide Web; see especially Sokal's site: http://physics.nyu.edu/faculty/sokal/.

10

Naturalistic Philosophy in Theory and Practice

10.1 What Is Naturalism?

What *kind* of theory should the philosophy of science try to develop? The logical empiricists had a definite answer to this question: the philosophy of science is concerned above all with the *logic* of science. By the middle of the 1970s, this view had well and truly broken down. Many wondered whether philosophy had become desiccated and irrelevant. As we saw in the previous chapter, this led to attempts by other fields to annex some of the traditional territory of philosophy of science. If philosophers could not say anything useful about how science works, others would do it instead.

Many philosophers came to agree that philosophy of science had to go beyond logical analysis, but there was less agreement on what should be done instead. In this chapter we look at one increasingly popular answer to this question: *naturalism*.

Naturalism is often summarized by saying that "philosophy should be continuous with science." This slogan sounds nice, but it is hard to work out what it really means. Naturalists reject the idea that philosophy should be sharply separated from other fields. In particular, naturalists hold that there should be some kind of close connection between scientific theories and philosophical theories, but they do not all agree on what this connection should be like. And what does a naturalistic outlook on philosophy mean in *practice*? Is it any more than a slogan and a label? In this chapter I will describe naturalism in general and then illustrate the naturalistic approach with an example. The next chapter will continue along the same lines. And from this point onward, the book starts to depart from the chronological structure that guided earlier chapters. The remainder of the book is organized more by topic than by chronology.

A moment ago I said that naturalists hold that philosophy should be continuous with science but do not agree on what this continuity is. Perhaps a more useful summary of naturalism is the idea that philosophy can *use* results from the sciences to help answer philosophical questions and can do this *even in the philosophy of science itself.*

From the perspective of many other philosophical positions, to *use* scientific ideas when theorizing *about* science involves a vicious circularity. How can we assume, at the outset, the reliability of the scientific ideas that we are trying to investigate and assess? Surely we have to stand *outside* of science when we are trying to describe its most general features and assess the integrity of its methods.

The idea that we should do the philosophy of science from an external and more secure standpoint is often referred to as *foundationalism.* (This term is sometimes used for other ideas as well.) Foundationalism requires that no assumptions be made about the accuracy of particular scientific ideas when doing philosophy of science. This is because before our philosophical theory is established, the status of scientific work is in doubt. One way to describe naturalism is to say that it is opposed to foundationalism in philosophy.

Naturalists think that the project of trying to give general philosophical foundations for science is always doomed to fail. They also think that a philosophical foundation is not something science needs in any case. Instead, we can only hope to develop an adequate description of how knowledge and science work if we draw on scientific ideas as we go. And the description of knowledge and science that results will be no more certain or secure than the scientific theories themselves.

Most philosophers who call themselves naturalists would agree with that sketch. From there on, however, there is a lot of disagreement. "Naturalism" is one of those words that a wide range of people find appealing as a label for themselves. As Elliott Sober likes to say, the term suggests that one's theories contain "no artificial ingredients." Philosophers, like shampoo manufacturers, would always like to call their products "natural." So there is a risk that naturalism as a movement will be swamped by the overuse of the term and will dissolve into platitudes. Despite this risk, "naturalism" is the label I use for most of my own philosophical work, and throughout the rest of this book I will often suggest that naturalism is our best hope for solving the core problems of philosophy of science.

10.2 Quine, Dewey, and Others

Where did contemporary naturalistic philosophy come from? The birth of modern naturalism is often said to be the publication of W. V. Quine's paper "Epistemology Naturalized" (1969). Certainly Quine's work is very important here, but we should not think of modern naturalism as coming entirely out of Quine. The American philosopher John Dewey is usually thought of as a pragmatist, but during the later part of his career (from roughly 1925 onward) his philosophy was a form of naturalism. In some areas Dewey's version of naturalism is superior to Quine's. But Dewey's philosophy was neglected during the second half of the twentieth century, and Quine is definitely the figure who had the most influence on naturalism. (Quine once acknowledged Dewey as an earlier naturalist, but Quine experts regard this as a polite gesture rather than a sign of real influence.)

Quine's article "Epistemology Naturalized" made a number of claims. He first attacked the idea that philosophers should give "foundations" for scientific knowledge. Quine's claims on this point have become central to naturalistic philosophy. But Quine also made a more radical claim. He suggested that epistemological questions are *so* closely tied to questions in scientific psychology that epistemology should not survive as a distinct field at all. Instead, epistemology should be absorbed into psychology. The only questions asked by epistemologists that have any real importance, in Quine's view, are questions best answered by psychology itself. Philosophers should expect that psychology will eventually give us a purely scientific description of how beliefs are formed and how they change, and we should ask for no more.

This version of naturalism is one that I, and many others, oppose. This opposition will probably come as no surprise; Quine seems to be claiming that philosophers interested in questions about belief and knowledge should close up shop and go home. Just as scientists warm to the heroic description of scientific work given by Popper, many philosophers turn a cold shoulder to Quine's claim that there is nothing important for us to do (unless we get psychology degrees and move to the psychology department). But aside from wanting to keep the paychecks coming, there is a deeper issue here.

In a different version of naturalism, there is such a thing as a *philosophical question,* distinct from the kinds of questions asked by scientists. A naturalist can think that science can contribute to the *answers* to philosophical questions, without thinking that science should replace philosophical questions with scientific questions. That is the version of naturalism that I defend. This contrasts with the kind of naturalism described by Quine in his 1969 paper; there we think of science as the only proper source of questions as well as the source of answers.

If we think that philosophical questions are important and also tend to differ from those asked by scientists, there is no reason to expect a replacement of epistemology by psychology and other sciences. Science is used as a resource for philosophy, not as a replacement.

What might be examples of these questions that remain relevant in naturalistic philosophy but which are not directly addressed within science itself? Many naturalists have argued that *normative* questions are important examples here—questions that involve a value judgment. If epistemology was absorbed by psychology, we might get a good description of how beliefs are actually formed, but apparently we would not be told which belief-forming mechanisms are good and which are bad. We would not be able to address the epistemological questions that have to do with how we *should* handle evidence, and how we can tell a good argument from a bad one. Those questions are central to philosophy. For the naturalist, the answers to these questions will often *depend* on facts about psychological mechanisms and the connections that exist between our minds and the world. But the naturalist expects that it will remain the task of philosophy to actually try to answer these questions. The sciences tend to concern themselves with different issues.

The term "normative naturalism" is often used for naturalistic views that want to retain the normative side of epistemology. (The term was coined by Larry Laudan [1987]; see also Kitcher 1992). I should also note here that although Quine's original discussions seemed to leave no place for normative questions in epistemology, toward the end of his career he modified his view, bringing it closer to normative naturalism (Quine 1990).

Normative naturalism accepts many (though not all) of the normative questions that have been passed down to us from traditional epistemology. But what is the basis for making these value judgments? What is the basis for a distinction between good and bad policies for forming beliefs? At this point normative naturalism confronts one aspect of the old and difficult problem of locating values in the world of facts. In the face of this problem, normative naturalists have often chosen a simple reply. The value judgments relevant to epistemology are made in an *instrumental* way.

In philosophical discussions of decision-making, an action is said to be *instrumentally rational* if it is a good way of achieving the goal that the agent is pursuing, whatever that goal might be. When assessing actions according to their instrumental rationality, we do not worry about where the goals come from or whether they are appropriate goals. We just ask whether the action is likely to achieve the outcome that the agent desires. And if some action *A* is being used as a means to *B*, it is a *factual* matter whether or not *A* is likely to lead the agent to *B*. So it is a factual matter whether or not action *A* is instrumentally rational for that agent.

It is uncontroversial to say that *one* kind of rationality is instrumental rationality. It is much more controversial to say that this is the *only* kind of rationality. Some normative naturalists think that instrumental rationality is the only kind of rationality that is relevant to epistemology. The problem of assessing which of a person's ultimate goals are justifiable is either rejected (because there can be no such assessment) or not addressed.

Earlier I said that John Dewey's work in the 1920s and 1930s describes and applies a good version of naturalism. Dewey's handling of the question of epistemological norms and values is one strength of the view. In his 1938 book *Logic,* Dewey develops a version of what would now be called normative naturalism. He says that in his epistemology, claims of "good" and "bad" reasoning are intended in the same way that we would understand claims about "good" and "bad" farming (1938, 103–4). Everyone is aware, he says, that some farming techniques are better than others at achieving the usual sorts of goals that farmers have. The likely consequences of different farming decisions are a factual matter, and we learn about these consequences from experience. The methods of farming that we presently regard as good ones are not perfect, and they might be improved further in the future. But there is no philosophical problem with making value judgments of this kind. Dewey says that the same is true of value judgments in epistemology.

I have focused on the role of normative questions here, but these are not the only kinds of questions that belong within epistemology, as opposed to the sciences that "feed into" epistemology. Another set of questions asked by philosophers have to do with the relationships between our common-sense or everyday view of the world, on one hand, and the scientific picture of the world, on the other. What kind of match (or mismatch) is there between the two pictures? We find questions of this kind in epistemology: what relationship is there between the common-sense or everyday picture of human knowledge, and a scientific description of our real contact with the world?

Answering this kind of question requires that we summarize both the everyday and the scientific pictures in a concise way and then compare them. One of the fastest-moving and most interesting parts of naturalistic philosophy in recent decades has been the naturalistic philosophy of mind. How does the everyday picture of the human mind and its contents (thoughts, beliefs, desires, memories) compare with the picture of the mind that is emerging from psychology and neurobiology?

Another set of philosophical questions that remain pressing for the naturalist have to do with the relations *between* different sciences. The various sciences each give us fragments, based on empirical work, of what the world is like and how it runs. But do the fragments tend to fit together neatly, or are there mismatches and tensions between them? The philosopher patrols the relationships between adjacent sciences, occasionally climbing into a helicopter to get a synoptic view of how all the pieces fit together. This can result in philosophical criticism of particular scientific ideas, but the criticism is made from the point of view of our overall scientific picture.

So I can now summarize the version of naturalism that I accept. Naturalism in philosophy requires that we begin our philosophical investigations from the standpoint provided by our best current scientific picture of human beings and their place in the universe. We begin from this picture, and we do not try to give a general justification, from outside of science, for our entitlement to use it. The science we rely on is not completely certain, of course, and may eventually change. The questions we try to answer, however, need not be derived from the sciences; our questions will often be rather traditional philosophical questions about the nature of belief, justification, and knowledge. Science is a *resource* for settling philosophical questions, rather than a replacement for philosophy or the source of philosophy's agenda.

I should note that I am not arguing that *all* work done by philosophers should be naturalistic in this sense. In particular, philosophy has long served another unusual and useful role in intellectual culture; it has acted as an "incubator" for novel, speculative ideas, giving them room to develop to a point where they may become scientifically useful. There are other roles for philosophy as a discipline as well. Philosophy often benefits, in fact, from its somewhat loose organization and open-ended agenda. We never know what new ideas, issues, or approaches might appear. But to the extent that we can expect to *solve* the big problems in fields like epistemology, naturalism is probably the right approach.

So what questions should we address in naturalistic philosophy of science? Back in chapter 1, I distinguished two sets of questions that would shape this book. We should try to achieve (1) a general understanding of how humans gain knowledge of the world around them, and (2) an understanding of what makes the work descended from the Scientific Revolution different from other kinds of investigation of the world. Those summaries are a start, but I can now fill in a bit more detail.

Does a naturalistic investigation of the role of observation in science support the familiar idea that observation and experiment make science *responsive* to the real structure of the world? One way to understand the work of some sociologists of science, including Latour, is to see it as proposing a theory of scientific change that gives *no* role to this notion of "responsiveness" to the world. Where exactly does a view like that go wrong? There *could,* in principle, be an institution that looked like what we call "science" but in which there was no genuine responsiveness to the world. Experiments would be no more than expensive "PR" exercises, and theories would change via a process of negotiation between factions. How do we know that our own science is not like this? In order to resolve this issue, we need to distinguish some "in principle" questions and some "in practice" questions. Does the nature of human thought and perception allow that, in principle, scientific belief can be made responsive to the real structure of the world? Even if this is possible in principle, do actual scientific communities operate in a way that makes this responsiveness occur in practice?

Suppose we are indeed able to vindicate the idea that science is responsive to the world; then what *sort* of contact with the world do our successful theories achieve? It is familiar to think of *truth* as the goal that we set for our theories; a good theory is one that represents the world truly. But is the traditional concept of truth a coherent and useful one here? Does it help us understand scientific progress at all? (Kuhn believed it does not.) What sense, if any, can a naturalist make of the idea of an "inductive logic," or a general theory of evidence? If we are able to isolate features of scientific thinking or scientific community structure that seem powerful and valuable, how can these be safeguarded and strengthened? Are there features of science that are self-defeating or harmful, which we might try to resist or change?

This version of naturalism will guide the remaining chapters of the book. But as I emphasized earlier, there are many different views that people like to label as naturalistic. And even philosophical discussions of science that are not carried out under the banner of naturalism have become more responsive to a variety of sources of ideas. This broadened perspective on the kinds of information that might be helpful has been a notable feature of recent philosophy. Some philosophers think the result has been chaotic, a profusion of ambiguous fragments and half-finished forays in too many directions. Discipline has been lost. But others, including me, think that the result has generally been progress.

10.3 The Theory-Ladenness of Observation

In this section I will focus on a debate that developed in the 1960s and continues to the present. The debate concerns the role of observation in science, and it is often called the debate about the "theory-ladenness of observation." Put most simply, the debate has to do with whether observational evidence can be considered an unbiased or neutral source of information when choosing between theories, or whether observations tend to be "contaminated" by theoretical assumptions in a way that prevents them from having this role. The problem is especially important for people who want to develop empiricist views. Advocates of radical theories of science, of the kind discussed in the last few chapters, have often seen the theory-ladenness of observation as a powerful argument against mainstream empiricism.

So the debate is important. The reason it is discussed in *this* chapter is that this debate becomes much easier to settle if it is approached from a naturalistic point of view. This issue gives us a good illustration of how naturalistic philosophy works in practice.

Our topic is observation. But "observation" is being understood in a very broad way here, to include all kinds of sensory contact with the world, all kinds of perception. Empiricists have agreed that observation is our source of knowledge about the world. Despite a good deal of disagreement within the empiricist movement, observation has generally been seen as *theory-neutral.* This neutrality, or absence of bias, is often the basis for the claim that observation is an "objective" way to settle disagreements.

It was against this background that arguments for the "theory-ladenness" of observation developed, especially in the work of N. R. Hanson, Kuhn, and Feyerabend. These arguments are a mixture, but their intended upshot is clear: observation cannot function as an unbiased way of testing theories (or larger units like paradigms), because observational judgments are affected by the theoretical beliefs of the observer. Therefore, traditional empiricist views about the role of observation in science are false.

As I said, these arguments are a mixture. Sometimes they are about the language of observation reports, sometimes about observation as a psychological phenomenon, sometimes about the beliefs resulting from observation, sometimes all of these. And while some of the phenomena discussed in the arguments are important and challenging, others are not. Some arguments only trouble logical positivism, while others trouble all possible views about science other than radical skepticism or extreme relativism.

Let us start with the more innocuous arguments. Sometimes it is claimed that observation is *guided* by theory, because theories tell scientists where to look and what to look for. This is true, but no sensible empiricist has ever denied it. This fact does not affect the capacity of observation to act as a test of theory, unless scientists are refusing to look where unfriendly observations might be found. All empiricists would regard that as a breakdown of fundamental scientific procedures.

At other times it is claimed that scientists must use theoretical assumptions to decide which observations to *take seriously.* Some apparent observations might involve malfunctions or mistakes of various kinds and can be disregarded. The observations that affect theory choice are "filtered" through a process in which some data are discarded. Because theoretical beliefs affect this filtering, there is the possibility of bias here.

Those problems are real. What they involve, however, is the problem of holism about testing, which was introduced back in chapter 2. Philosophers are still trying to unravel this problem, as they try to develop new theories of testing and confirmation. In the absence of a general solution to these problems, some pieces of an answer can be suggested. The theoretical assumptions that affect the relevance of an observation to a piece of theory can themselves be tested separately. We might also venture some low-level recommendations: perhaps in crucial tests, scientists should be more reluctant to discard observations. But in this area it is hard to know which pieces of common sense are helpful, which are trivial, and which are flat wrong.

Another set of arguments about observation concern language. When a scientist has an experience, he or she can only make this experience relevant to science by putting it into *words*. The vocabulary used, and the meanings of even innocent-looking terms, will be influenced by the scientist's theoretical framework. Given the interconnections between the meanings of words in a language, there is no part of language whose application to phenomena is totally "theory-free."

Some versions of this argument are not of enduring importance because they cause trouble only for the logical positivist ideal of a purely observational language, sharply distinct from the parts of language that use or assume theoretical ideas. Sometimes critics of empiricism write as if once it has been shown that the language of observation is in some sense "theoretical," that is the end of the argument and empiricism is dead. This is a mistake. In working out the relevance of this issue to more modern forms of empiricism, everything depends on which *kinds* of theories affect the language of observation and on the *nature* of this effect. For example, maybe observational reports assume "theories" that are so low-level that the testing of real scientific theories will never be affected. We can think of the assumption that objects generally retain their shape when we are not looking at them as "theoretical" in a sense, but the effect of this assumption on observation reports does not usually matter to testing in science.

But suppose it can be shown that observation reports are affected by the kinds of theories that are themselves being tested. For example, Feyerabend tried to show that innocent-looking descriptions of motion in the seventeenth century were affected by theoretical background assumptions in this way. This looks like trouble. But even this kind of effect may or may not be philosophically important. A theory might contribute the *concepts* used to express an observation, without this affecting the capacity of an observation report to test the theory in question. Not every result described in terms of the *concepts* preferred by theory *T* will be an observation report that is *favorable* to theory *T*. Back in my discussion of Popper, I mentioned that an observation of rabbit fossils in Precambrian rocks would be a massive shock to evolutionary theory (section 4.6). Suppose we regard "I saw rabbit fossils in Precambrian rocks" as an observation report that is very much "laden" with biological and geological theory. Some might want to say it is so laden with theory that it is not an observation report at all. But regardless of this, the report would *still* be a massive shock to evolutionary theory.

So imagine that we had a simple falsificationist view of testing in science. It is clear that the fact that observation reports are expressed using concepts derived from a theory has no effect on the capacity of nature to say NO to a conjecture. Simple falsificationism is not an adequate view of testing in science, but that does not matter to the present point. The point is that an influence of theory on observational vocabulary does not, on its own, prevent observation from acting as an unbiased test of theory.

The final aspect of the theory-ladenness arguments that I will consider is the most important. Kuhn and others have argued that even the *experiences themselves* that a person has are influenced by their beliefs, including their theories. Not just the use of observation reports in assessing theories, and not just the linguistic form of the reports are affected, but the perceptual experiences themselves are. There is *no stage* in the processes of observation in science where theories do not play a role.

When giving these arguments, Kuhn and others liked to use the results of psychological research in the middle of the twentieth century. This research was taken to refute a "passive" view of perception and replace it with a view holding that perception is active and intelligent. Psychologists emphasized the multiple ways in which a pattern of stimulation on the retina could be caused by objects in the world. If there are multiple possibilities, then theoretical assumptions must be used by the visual system to make a choice (Gregory 1970).

This kind of theory-ladenness argument was attacked by Jerry Fodor in a very convincing (and very funny) 1984 article called "Observation Reconsidered." My reply to the argument largely follows his. As in the case of the influence of theory on observational language, everything hinges on *which* theories affect observation and *how* they affect it.

Fodor turned the tables on some theory-ladenness arguments via a discussion of perceptual illusions of the kind often discussed in psychology textbooks. Consider the Müller-Lyer illusion, represented in figure 10.1. The two lines are the same length, although we tend to see the lower one as longer than the upper one. According to psychology, the illusion is brought about by the unconscious use of background assumptions in the processing of visual inputs. People have taken this result to show a kind of theory-

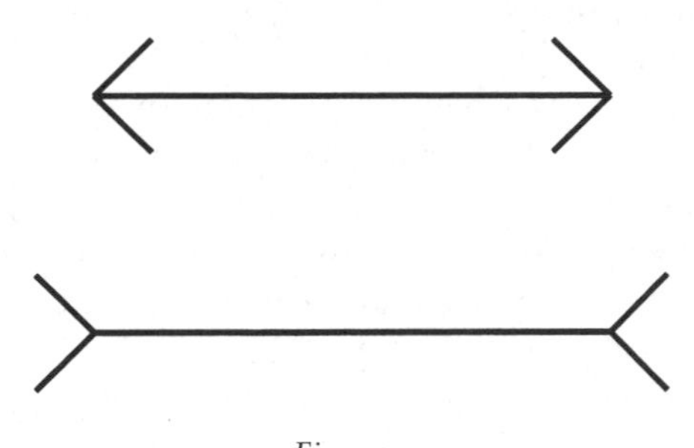

Fig. 10.1
The Müller-Lyer illusion

ladenness in perception; we do not realize it, but our general and implicit beliefs about the world are affecting what we see.

But Fodor then notes that although it is true that the illusions seem to be produced by the effects of unconscious theory, *some* pieces of theory or background knowledge seem to have *no* effect on perception. Most notably, the illusion is not affected *by the knowledge that it is an illusion,* or by knowledge of the theory of illusions. Those pieces of background knowledge do not make the illusion go away. We seem to have a situation in which the mechanisms of perception are influenced by some theories and not others. And the theories that do have this effect are not high-level scientific theories. They are low-level sets of assumptions about the physical layout of the world—the three-dimensional nature of space, the effect of distance on apparent size, and so on.

Fodor links his argument to a research program in psychology that posits *modules* in the explanation of perception and some other tasks (Fodor 1983). Modules are automatic, innate pieces of mental machinery that do their processing unconsciously and make use of a fixed *subset* of a person's background knowledge. In perception, modules send their output to the "central" cognitive mechanisms. These central mechanisms have access to all a person's theories and ideas when working out what to *do* with the observation. So although the later stages of responding to observation are affected, in principle, by all the theories a person might have, the output of the perceptual module is not. The module's operation—which determines how things *seem* to a person—is not biased by whether the person accepts one scientific theory or another.

As Fodor says, this does not solve the entire problem of the role of observation in testing. Even if observation itself is not biased by commitments to scientific theories, as has been alleged, there is still the issue of what a person *does* with the observation. That takes us back to the problem of holism about testing.

Not all the problems have been solved, but we have made progress. And I think it is clear how this discussion provides support for a naturalistic approach to philosophy of science. Observation is a natural phenomenon, studied by fields like psychology and psychophysics. Those disciplines tell us what perceptual mechanisms are like and what kind of connection we have to the world via these mechanisms. Naturalistic philosophers can put these results to use in working out how observation operates in science generally.

There are two general sets of questions about observation that we need to answer.

1. To what extent is observation a *reliable* way of forming true beliefs about what is going on in the world? When is observation using ordinary human senses reliable, and when is it not?
2. Is observation *neutral* between competing theories of the kind we wish to test in science? Can people with very different theoretical commitments agree about what has been observed? That is, does observation provide an *intersubjective* basis for theory choice?

These two sets of issues are distinct, but they are connected in complicated ways. For example, if we have reason to think that observation using ordinary human senses is very reliable in normal conditions, then we can expect a wide range of people to *agree* when they are looking at the same thing in normal conditions. This claim could be hedged more, but the basic idea is clear. Reliable senses shared across all normal humans can be expected to deliver consensus. But it is also possible that we could have wide agreement without reliability; we might all be deluded in the same way. Some philosophers have thought that color vision is like this. Colors are not really "out there" in the world, even if we all experience them that way.

In order to assess whether observation is neutral between competing theories, the kind of evidence that Fodor presents in his 1984 paper is relevant. Although his arguments are rather convincing, the important thing here is not whether a "modular" theory of perception is ultimately right. The important thing is to see what *kind* of scientific evidence is relevant for settling the question. In order to work out whether observation in scientific communities is affected by theoretical differences in a way that threatens empiricism, we need to work out what human perceptual mechanisms are like, whether they tend to be similar in all normal humans, and what role high-level scientific beliefs have in the process of observation. This kind of evidence does not itself settle whether observation is a *reliable* way of forming true beliefs about what is going on in the world. But that is something that can be investigated systematically by psychology and psychophysics.

Note that our perceptual mechanisms might use low-level theories in a way that makes perception reliable without the low-level theories *themselves* being true. What we are trying to assess is the reliability of observation reports themselves. We might have built into our brains something like Euclidean geometry, which is not strictly true of our universe, but which is used in such a way that we almost always end up with true observational judgments. Note that the Euclidean habits we have for interpreting space did not prevent us from revising the scientific belief that space strictly fits Euclidean geometry, as Einstein did in his theory of General Relativity.

I will finish this section by starting to sketch a version of empiricism that follows a naturalistic approach to the role of observation. This will be a recurring theme in the rest of the book.

We ask, What is the role of observation in science? To answer this, we need first to understand the actual role of observation in the sociological patterns of scientific activity. How is observation used as a resource in science? How is it used in the settlement of controversies? Then we can start to feed in results from the scientific study of observation. We ask, Given the kind of connection to the world that observation provides, and given the role of observation in science, what kind of contact does science itself have with the world? If observation is the channel by which theory makes contact with reality, what kind of channel is it? That is a question that we can only answer by drawing on the empirical sciences that deal with observation and perception.

If empiricism about science is vindicated by the answers to these questions, it will be a form of empiricism that differs from traditional forms. Observation is a form of physical contact between our minds and the world. This contact is the product of evolution, and it has whatever degree of reliability it has because of our evolutionary history and the contingent relationships between our structure and that of our surroundings. Science is an attempt to *exploit* this contact between our minds and the world, and science is also *motivated* by the limitations that result from our relations to the world; we need science because much of the world is not accessible to ordinary observation. Science works by taking theoretical ideas and *trying to find ways to expose them* to observation. The scientific strategy is to construe ideas, to embed them in surrounding conceptual frameworks, and to develop them, in such a way that this exposure is possible even in the case of the most general and ambitious hypotheses about the universe. That view is a kind of empiricism, and I think we can be optimistic about that kind of empiricism. This is a form of empiricism in which naturalism is, in a sense, primary. The advantages of empiricist philosophical ideas are not shown or established by philosophy alone.

Further Reading

The history of naturalistic philosophy is discussed in Kitcher 1992. Kornblith's *Naturalizing Epistemology* (1994) is a good collection of papers on the topic, including Quine's classics. Dewey's most important naturalistic work is his *Experience and Nature* (1929). Callebaut's *Taking the Naturalistic Turn* (1993) is an unusual exploration of naturalism based on interviews; his idea of who is a naturalist is also a bit unusual in some cases. For normative naturalism, see Laudan 1987. For good examples of naturalistic work in the philosophy of mind, see Dennett 1978, Fodor 1981, Stich 1983, and Dretske 1988.

Fodor's 1984 paper on observation is the subject of a response in the journal *Philosophy of Science* by Paul Churchland (1988). Fodor also replied in the same issue. See also Bishop 1992. See Giere 1988 for a treatment of various other issues in philosophy of science using a "cognitive" approach that draws on psychology.

11

Naturalism and the Social Structure of Science

What is good for General Motors is not always good for the nation, but once science is properly understood, it turns out that what is good for the individual scientist is by and large good for science.

DAVID HULL, *Science as a Process*

11.1 Science as a Process

Traditional empiricism neglected the social structure of science; naturalistic philosophy has tried to avoid this mistake.

David Hull's sprawling *Science as a Process* (1988), the source of the epigraph above, is the product of decades that Hull spent observing and interacting with biologists, especially the community of biologists who study *systematics*, the classification of organisms.

Hull's story begins with a fairly common-sense picture of both science and scientists. Scientists are curious about the world, as many other people are. Individual scientists make their way into a social structure that puts this curiosity to work and does so via empirical testing. So far this does not tell us much. But Hull then argues that what makes science special is an unusually good relationship between the motivations and goals of the individual scientist, on one hand, and the goals of science as a whole, on the other.

Is science a fundamentally cooperative enterprise, or is it a fundamentally competitive one in which scientists are out for personal advancement? According to Hull (and also Merton), science runs on a *combination* of cooperation and competition. Neither is fundamental, and the special features of science are due to an interaction between the two. This interaction arises from the reward system found in science and the context in which the reward system operates.

Hull argues that the main professional motivation for individual scientists is the desire for recognition. And one kind of recognition is most relevant: *use.* Scientists want other scientists to use their work, giving credit when they do so. Clearly Hull is close to Merton here, as he acknowledges. As we saw in chapter 8, Merton argued that recognition is the basic reward in science. There might be some differences, however. Hull stresses the desire to have one's ideas *used;* Merton stresses being recognized as the *first* to come up with an idea. Often these will go together, but not always. In Hull's view but not Merton's, if a scientist's version of an idea is used because it is especially convenient, even though it was not the first, that counts as a real reward. For Hull, being used and cited matters more than anything else.

What generates the distinctive features of science, as Hull sees it, is this reward system *operating in a special context.* Each scientist inherits the ideas and methods of her or his field from earlier workers. Of course, an individual might revolutionize the field, but even revolutionary work starts out from an inherited context. Individual scientists cannot do anything significant without entering a system of cooperation and trust. You can't make a contribution of your own without using the work of others. And in order to use the work of others in ways that provide support for what you are doing, you need to give citations. So the desire to do work that is used requires using, and citing, the work of others. Scientists trade credit for support, and they do so in the hope that others will do the same for them. This reciprocation is not primarily a matter of goodwill (although that can be a factor); it results from a special kind of self-interest.

Traditional descriptions of science often stress the *replicability* of results. A result is not trustworthy if only one lab seems able to produce it. But as Hull emphasizes, no one has the time to do much of this replication and checking. The checking that actually occurs is a consequence of the desire to have one's work used. In order to do work that others will use, you need to ensure that the work that *you* rely on can be trusted. So those who check and attempt to replicate results will often be those who need to know whether they can rely on those ideas. Another kind of person who will be motivated to engage in rigorous checking will be someone whose own work is either discredited or made less important by a piece of new work. Hull also argues that the reason why fraud in science is so much more serious a crime than theft, even in cases where public well-being is not affected, has to do with these sorts of factors. In a case of theft or plagiarism, the only person harmed is the one stolen from. But when a case of fraud is discovered, all the scientists who used the fraudulent work will find their work on that topic deemed unreliable, and their work will not be used.

So the desire to have one's work used generates many other features of science—the elaborate networks of citation, the real but selective checking of others' results, the relative seriousness of fraud and theft. And Hull argues that although this system malfunctions on occasion, the general result is a harmonious relationship between the behavior of individual workers and the goals of science as a whole. Cooperation and teamwork is common. Hypotheses are scrutinized closely. Idle speculation and shoddy work are discouraged. Ideas are shared freely (though not before they are in publishable form). Work produced by those who are low on the totem pole is taken seriously (especially by those whose own projects might be helped or harmed).

A number of philosophers and scientists have been attracted to a picture of science as a dialogue between an imaginative voice and a critical voice, between the speculative and the hard-headed. Popper is an example. This is an appealing picture, but why should this dialogue actually *occur*? Hull aims to give us a mechanism. And a key part of the mechanism is the distribution of roles across different people. In contrast to Popper, Hull argues that there is no need for individual scientists to take a cautious and skeptical attitude toward their *own* work; others will do this for them.

Back in chapter 6, I discussed the "invisible hand" structure of Kuhn's account of science. I made an analogy to Adam Smith's famous defense of market economies, which argued that a collection of selfish individuals interacting in a market tends to produce a good outcome for everyone. Kuhn's analysis of science sees a degree of individual-level dogmatism as contributing to openness at the level of science as a whole. In Smith and Kuhn we have a surprising explanatory relationship between individual-level properties and the properties of the whole. Hull's picture of science has some similarity to these cases, but as he says, the hand is not really hidden or "invisible" in the story he tells. The relationship between individual and group-level properties is not so unexpected.

So far we have looked at the consequences of the structure of reward and motivation in science. But *why* do scientists want to have their work used in the way Hull describes? This question is related to another: why has the social structure that Hull describes arisen so rarely, across the range of different communities that have wanted to understand the world?

Hull says that he assumes that both curiosity and a desire for recognition are fairly basic human motivations. But, I suggest, a desire for the specific kind of credit found in science is surely more unusual. What we seem to have is a situation in which a basic human desire for credit is shaped by the internal culture of science into a very specific desire for recognition in the form of *use*. More accurately, we should expect both some shaping and some selection here; individuals who do not find the scientific reward system satisfying might never finish graduate school.

Hull and Merton both look at possible roles for broader cultural features here. Modern science developed in European societies that were comfortable with the ideas of individual competition and credit. The reward system found in science was a fairly early invention. When the French Academy was founded in the seventeenth century, its members initially tried to handle credit in a communal way, but this did not work, so they switched quickly to a more individualistic approach. The Royal Society of London, under its skillful first secretary, Henry Oldenburg, used rapid publication in the *Proceedings* to allocate credit and to encourage people to share their ideas. Oldenburg's system, which also included anonymous refereeing of papers, is basically what has come down to us today.

In this outline of Hull's ideas, I have not yet mentioned one of his main arguments. Hull tries to describe scientific change as an *evolutionary* process, in a sense derived from biology. Science changes via processes of variation and selection, just as biological populations do. Individual ideas in science are replicated in something like the way that genes are. And the different rates with which ideas are replicated are consequences of their manifestations in the brains and the public representational systems (books, journals, computers) of the scientific community. Scientific change is a process in which some ideas outcompete others in a struggle for replication.

The idea of understanding scientific change via an explicit analogy with biological processes of variation and selection has been tried out by a number of writers (Toulmin 1972; Campbell 1974; Dennett 1995). As we saw in chapter 4, Popper's view of science also has an analogy with evolution by natural selection, although Popper did not start out with this analogy in mind.

Though the analogy between science and Darwinian evolution is something that people keep coming back to, the analogy has not yielded a lot of new insights so far. We find the same result in many other attempts to describe cultural change as an evolutionary system; a wide variety of processes *can* be described in a way that borrows from evolutionary biology, but usually this exercise does not teach us anything about those cultural processes that we did not know before. Biological populations have special features that make the abstract concepts of evolutionary theory *helpful* in trying to understand them. Other systems, which lack these features, can be described in evolutionary terms with a bit of shoehorning, but we do not seem to gain much from doing so.

That does not keep the analogy between evolution and scientific change from being an interesting one. The analogy can be interesting without being a basis for a new theory of science.

11.2 Kitcher and the Division of Scientific Labor

I move to a second example of naturalistic work on the social structure of science, this one from the work of Philip Kitcher.

In chapter 7 I discussed Lakatos's and Laudan's views about competition between research programs. Both presented a picture of science based on competition between teams of workers developing rival theories and perhaps defending rival methods. This picture seems to cover some parts of science rather well. Both Lakatos and Laudan were interested in giving normative rules for scientific behavior in this situation. But as I said back then, there is a gap in their treatment of the issue. They were thinking of rational choices by *individuals*. We can also look at the situation from the point of view of the scientific community, and we can ask, What is the best *distribution* of workers across rival research programs, for the community as a whole?

Kitcher takes up this issue in detail (1990, 1993). He starts by asking this question: suppose you ruled science "from above" and had to allocate resources to rival research programs. In a particular scientific field, you find two different approaches being taken to the same problem. Research program 1 looks more promising than research program 2, but no one knows which approach will ultimately work. However, it is clear that either one will succeed while the other fails, or both will fail. How should you allocate resources, to maximize the chance that the scientific problem will be solved?

The answer will depend on the details of the case, obviously. But it seems clear that in a wide variety of situations, the best approach will *not* be to allocate all resources to one option and none to the other. Some degree of "bet-hedging" will often be advisable, even in cases where one program is obviously more promising than the other. A wise "ruler of science" would often allocate most of the resources to the better research program but some resources to the alternative.

To say more than this, we need to represent the situation mathematically, and that is what Kitcher does. The crucial features of the situation will be the degree to which one program is more promising than the other and the mathematical functions that describe how each research program responds to the addition of more resources. Here is a simple case. Suppose that both research programs become more and more likely to succeed as more workers are added to them, but in both cases there is a "decreasing marginal return." As more workers are added to a program, each additional worker makes less and less difference to the chances of success. We can then see why an optimal allocation of resources will often not put every worker on one program. After a certain point, adding more workers to a program has almost no effect, and these people would be better put to work on the alternative. Unless the total pool of workers is small and the overall difference in promise is big, the best distribution of workers will allocate some to one program and some to the other.

That is what we would want if the allocation of resources could be controlled from above. But of course, this is not how things usually work. Now suppose that individual scientists are making their own choices about which program to work on. The next question Kitcher asks is, What kind of individual reward system in science will tend to produce distributions of workers that benefit science as a whole? What kind of reward system will tend to produce the *same* distribution of workers that the "ruler from above" would want?

One option that would *not* work well would be to give a fixed reward to everyone who works on the program that eventually succeeds, regardless of how many workers there are. That system would induce everyone to choose the more promising program, and the community would have all its eggs in one basket. Another approach would be to reward individuals for making choices that produce the maximum benefit in terms of the overall chance that the community will solve the problem. This would work in principle, but it does not seem a realistic reward system for actual scientific communities. So here is a third option: we reward only the individuals who work on the research program that succeeds, but we divide the "pie" equally between all the workers who chose that program. So the reward that an individual gets will depend not just on their own choice but on how many other individuals chose the same program.

This third reward system, Kitcher argues, will produce a good distribution of workers across the two options. We can see why that is. Once one research program becomes crowded, an individual has little incentive to join that program because, if it does pay off, the pie is being divided among too many people. Although the other program is less likely to succeed, if it *does* succeed, there will be fewer workers sharing the reward. So an individual who wants to maximize his or her "expected payoff" will often have reason to choose the less promising program. In this way, selfish individual choices will produce a good outcome for the whole community. And Kitcher suggests that this reward system is fairly close (with simplifications) to what we actually find in science. The "pie" here is not cash but prestige.

Kitcher's story has a definite "invisible hand" structure. We have selfish individual behaviors combining to produce a good outcome for the community. This outcome might be one that the individuals are uninterested in or even unaware of.

Kitcher's work has recently been followed up by Michael Strevens (2003). Strevens shows that Kitcher was too optimistic about the reward system in which a fixed pie is shared equally by all workers on a successful program. Although this reward scheme will tend to produce a fairly good distribution of workers from the point of view of the community, it often will not produce the best distribution. Suppose you are making the choice of which program to join. There are cases where you will do best to join the more promising program even though your joining makes *little or no difference* to its chance of success. Others have given the program a good chance of success, and your joining gives you a good chance of an equal share of the pie, though your efforts would have been more productive if you had joined the alternative program. Had you joined the alternative, you could have made a real difference to the community's overall chance of solving the problem. So a kind of "free riding" is encouraged in Kitcher's reward scheme.

Strevens argues that another reward scheme is both better for the community *and* closer to the actual situation in science. This scheme allocates rewards to an individual that are proportional to the contribution he makes to the particular research program that he joins. The payoff is given only if a research program solves the scientific problem, and the pie is shared unequally among those working on the successful program. Workers who joined early and made a big difference to the program's chance of success get more than workers who joined late and made little difference.

There is obviously a great deal more detail that could be added here; I have just introduced the simplest part of a complicated model. The overall picture is clear, though. Hull, Kitcher, Strevens, and others are looking at the relationship between individual incentive and community-level success in science. The argument—made most bluntly by Hull but endorsed by others—is that science has hit on a particularly effective way of coordinating individual energies to yield good outcomes for the community as a whole.

11.3 Social Structure and Empiricism

At the end of the previous chapter, I began to sketch a version of empiricism based on a naturalistic approach to philosophy. Science attempts to exploit the contact with the world that humans have via experience, using this contact to explore and assess hypotheses about the world. In this sense, we might think of science as a *strategy* for answering questions and working out what to believe. To a very limited extent, this strategy can be followed by a lone individual, but what results is something that has few of the distinctive features of science. The power of science is seen in the *cumulative and coordinated* nature of scientific work; each generation in science builds on the work of workers who came before, and each generation organizes its energies via collaboration and public discussion. This social organization permits the scientific strategy to function at the level of social groups; the dialogue between the speculative voice and the critical voice can literally be a dialogue, rather than something internalized in the mind-set of the individual scientist. These social groups can include some individuals who are not especially open-minded—who are very wedded to their own ideas—provided that the group as a whole retains flexibility and responsiveness to evidence.

So how do we get a community of individuals to behave in this sort of way? We need a suitable reward system and also various external supports. Some of what is needed is obvious; scientists need to be able to make a living, unless we intend to leave it all to rich amateurs. The society as a whole must allow questioning and open-ended inquiry. Though these factors are obvious, other needs are probably more subtle. The work of people like Merton and Hull suggests that science may need a specific kind of internal culture and reward system; the delicate balance between competition and cooperation is not easily achieved. But there are many unanswered questions here. Might science work just as well, or better, with a slightly different reward system from what we find today? Do we really need the intense and often egoistic competition found in the science of Western market-based societies? Those who like competitive, individualist societies will be inclined to say *yes;* they will think that nothing else can generate the precious patterns of scientific social behavior. Those who dislike individualism and competition, who prefer a more communitarian or socialist society, might say *no;* they will hope that we could do as well or better with a different reward system and a less competitive atmosphere.

This is a point where some feminist discussions of science are relevant. Some feminists hold that the competitive and individualist culture of science is more in tune with the temperaments of men than of women. This affects the ability of many women to flourish within the culture of mainstream science. If it is real, this exclusion may have epistemological consequences. Suppose it is also true that women bring a different "style" of thought and investigation to science and that science benefits from diversity of this kind. In that case (and there are a *lot* of "ifs" here), the competitive culture of science will tend to produce subtle kinds of uniformity in scientific thinking and will reduce the frequency of a valuable kind of input into scientific discussion.

So we see that there are several ways in which Hull's blunt assertion of the harmony between individual-level and group-level benefit in science might be overstated. One of Hull's own examples is interesting here. Hull discusses, and extends, work by sociologists on the temperament and leadership styles of successful scientists. The data suggest that an aggressive faith in one's own ideas, the pushiness of a "true believer," can be useful in at least some fields. Hull refers to a sociological study in which a detailed survey of students and colleagues was used to investigate the temperaments of some famous twentieth-century psychologists in the United States. One contrast is especially interesting, that between B. F. Skinner and E. C. Tolman. Both Skinner and Tolman were in the "behaviorist" tradition in psychology; they wanted psychology to be experimental, quantitative, and closely focused on behavior. But Skinner's version of this approach was almost absurdly strict, while Tolman's was more flexible. Tolman was also a modest, open-minded, considerate sort of person; Skinner was dogmatic and pushy. And Skinner, with his crusading zeal, had much more influence than Tolman. We cannot know for sure what role temperament had in explaining this difference in success, of course, but the data are suggestive (and Hull found similar results in a smaller study of his own).

So suppose that pushiness and zeal work well for individuals. Does this tend to result in good outcomes for science? In this case, a good argument could be made for the opposite view. I conjecture, and many psychologists would agree, that if Tolman had dominated mid-twentieth-century psychology rather than Skinner, it would have been far better for the field. (Some of Tolman's ideas are currently being revived [Roberts 1998].) I suspect that Hull might reply that the possible problems raised by the feminist objection just above, and those illustrated by the case of Skinner and Tolman, are a small price to pay for the benefits gained from the present balance of competition and cooperation in science.

We must also bear in mind that the internal culture of science is not something fixed and unchangeable. The ideas of people like Merton, Kuhn, Hull, and Kitcher might describe science from the seventeenth to the twentieth centuries, but change may well be in the air (Ziman 2000). Scientists have usually not hoped to become rich through their work; recognition, especially by their peers, has been an alternative form of reward. But a number of commentators have noted that big financial rewards have now started to become a far more visible feature of the life of the scientist, especially in areas like biotechnology. Kuhn warned that the insulation of science from pushes and pulls deriving from external political and economic life was a key source of science's strength. We do not know how fragile the social structure of science might be.

In any case, this chapter and the previous one have introduced some of the main themes in naturalistic philosophy of science. Naturalists hope that by combining philosophical analysis with input from other disciplines, we will eventually get a complete picture of how science works and what sort of connection it gives us to the world. This last issue—the connection that science gives us to the real world we inhabit—has often been mishandled by sociology of science and Science Studies. That is the topic of the next chapter.

Further Reading

For assessments of Hull's theory of science, see the reviews in *Biology and Philosophy,* volume 3 (1988). Also see Sterelny 1994.

Kitcher's main work here is *The Advancement of Science* (1993). The model discussed in this chapter is presented in a simpler form in Kitcher 1990. Solomon (2001) defends a "social empiricism" in detail, with many examples from the history of science.

For a more general discussion of social structure and epistemology, see Goldman 1999. Downes (1993) argues that some naturalists do not take the social nature of science seriously enough. Sulloway 1996 is a very adventurous discussion of the role of personality and temperament in scientific revolutions.

12

Scientific Realism

12.1 Strange Debates

What does science try to describe? The world, of course. Which world is that? *Our* world, the world we all live in and interact with. Unless we have made some very surprising mistakes in our current science, the world we now live in is a world of electrons, chemical elements, and genes, among other things. Was the world of one thousand years ago a world of electrons, chemical elements, and genes? Yes, although nobody knew it back then.

But the concept of an electron is the *product* of debates and experiments that took place in a specific historical context. If someone said the word "electron" in 1000 A.D., it would have meant nothing—or at least certainly not what it means now. So how can we say that the world of 1000 A.D. was a world *of* electrons? We cannot; we must instead regard the existence of electrons as dependent on our conceptualization of the world.

Those two paragraphs summarize one part of an argument about science that has gone on constantly for the last fifty years, and which stretches much deeper into the history of philosophy. For some people, the claims made in the first paragraph are so obvious that only a tremendously confused person could deny them. The world is one thing, and our ideas about it are another! For other people, the arguments in the second paragraph show that there is something badly wrong with the simple-looking claims in the first paragraph. The idea that our theories describe a real world that exists wholly independently of thought and perception is a mistake, a naive philosophical view linked to other mistakes about the history of science and the place of science in society.

These problems have arisen several times in this book. In chapter 6 we looked at Kuhn's claim that when paradigms change, the world changes too. In chapter 8 we found Latour suggesting that nature is the "product" of the settlement of scientific controversy. I criticized those claims, but now it is time to give a more detailed account of how theory and reality are connected.

12.2 Approaching Scientific Realism

The position defended in this book is a version of *scientific realism.* A scientific realist thinks it *does* make sense to say that science aims at describing the real structure of the world we live in. Does the scientific realist think that science *succeeds* in this aim? That is a more complicated issue.

Formulating scientific realism in a precise way will take a while. And the best way to start is to ignore science for the moment and look first for a more general description of "realist" attitudes.

The term "realism" gets used in a huge variety of ways in philosophy; this is a term to be very cautious about. One tradition of dispute has to do with what our basic attitude should be toward the world that we seem to inhabit. The simple, common-sense view is that the world is out there around us, existing regardless of what we think about it. But this simple idea has been challenged over and over again. One line of argument holds that we could never *know* anything about a world of that kind. This debate has carried over into the philosophy of science.

How might we give a more precise formulation of the "common-sense" realist position? The usual starting point is the idea that reality is "independent" of thought and language (Devitt 1997). This idea is on the right track, but it has to be understood carefully. People's thoughts and words are, of course, real *parts* of the world, not extra things floating somehow above it. And thought and language have a crucial *causal* role in the world. One of the main reasons for thinking, talking, and theorizing is to work out how to affect and transform things around us. Every bridge or light bulb is an example of this phenomenon. So a realist statement about the independence of the world from thought must have some qualifications. Here is my formulation:

Common-sense Realism: We all inhabit a common reality, which has a structure that exists independently of what people think and say about it, except insofar as reality is comprised of, or is causally affected by, thoughts, theories, and other symbols.

The realist accepts that we may all have different *views* about the world and different perspectives on it. Despite that, we are all here living in and interacting with the same world. Let us now return to issues involving science.

12.3 A Statement of Scientific Realism

How should scientific realism be formulated? One possibility is to see the scientific realist as asserting that the world really is the way it is described by our best-established scientific theories. We might say: there really are electrons, chemical elements, genes, and so on. The world as described by science is the real world. Michael Devitt is an example of a scientific realist who expresses his position in this way (1997).

My approach will be different. I agree with Bas van Fraassen, and others, who argue that it is a mistake to express the scientific realist position in a way that depends on the accuracy of our current scientific theories. If we express scientific realism by asserting the real existence of the entities recognized by science now, then if our current theories turn out to be false, scientific realism will be false too.

Should we worry about the possibility that our best-established theories will turn out to be wrong? Devitt thinks that so long as we do not commit ourselves to realism about speculative ideas at the frontiers of science, we need not worry. Others think that this confidence shows disregard for the historical record; we should always recognize the genuine possibility that well-established parts of science will run into trouble in the future.

How do we decide this massive question about the right level of confidence to have in current science? My suggestion is that we *don't* decide it here. Instead we should separate this question from the question of scientific realism. A scientific realist position is compatible with a variety of different attitudes about the reliability of our current theories. We want a formulation of scientific realism that is expressed as a claim about the enterprise of science as a whole.

One complication comes from the following question: must the scientific realist also be a common-sense realist? Is it possible—in principle—that science could tell us that common-sense realism is false? The problem is made vivid by puzzles with quantum mechanics, one of the basic theories in modern physics. According to quantum mechanics, the state of a physical system is partially determined by the act of measurement. Some interpretations of quantum mechanics see this as causing problems for common-sense realist ideas about the relation between human thought and physical reality. These interpretations of quantum mechanics are very controversial. Like a lot of other philosophers, I have been quietly hoping that further work will eventually show them to be completely mistaken. But that is not the point that matters here. The point is this: should we allow for the *possibility* that science could conflict with common-sense realism? If we say that scientific realism does assume common-sense realism, we seem to be committed to holding on to an everyday, unreflective picture of the world, regardless of what science ends up saying. But if we sever scientific realism from common-sense realism, it becomes hard to formulate a general claim about how the aim of science is to describe the real world.

My response to the problem is to modify common-sense realism so that it allows for the possibility of unexpected, uncommonsensical relations between thought and reality at large. Common-sense realism as previously formulated allowed for the possibility of *causal* links between thought and the rest of reality. It is often hard to tell whether a connection posited by science is a causal connection or not. So let us widen the class of relations between thought and the world that realism accepts; science might add new cases. Because we are modifying common-sense realism to make it more responsive to science, this is a *naturalistic* modification.

Common-sense Realism Naturalized: We all inhabit a common reality, which has a structure that exists independently of what people think and say about it, except insofar as reality is comprised of thoughts, theories, and other symbols, and except insofar as reality is dependent on thoughts, theories, and other symbols in ways that might be uncovered by science.

Once we have made this modification, it is reasonable to include common-sense realism as part of scientific realism. Here is my preferred statement of scientific realism:

Scientific Realism:

1. Common-sense realism naturalized.
2. One actual and reasonable aim of science is to give us accurate descriptions (and other representations) of what reality is like. This project includes giving us accurate representations of aspects of reality that are unobservable.

In this sense, I am a scientific realist.

Several comments on this formulation are needed. First, clause 2 says that *one* aim of science is to represent the structure of the world. Nothing implies here that this is the only aim of science. There might be other aims as well. And some particular theories—even whole research programs—might be developed in a way intended to serve other purposes.

Second, I said "actual and reasonable aim." The first part of this is a claim about the goals behind at least a good proportion of actual scientific work. The second part claims that scientists are not deluded or irrational in making this their goal. They can reasonably hope to succeed at least some of the time.

But I do not say how often they succeed. No part of my statement of scientific realism endorses our current particular scientific theories. In some areas of science, it's hard to imagine that we could be badly wrong in our current views; it's hard to imagine that we could be wrong in believing that tuberculosis is caused by a bacterium and that chemical bonding occurs via the interactions of outer-shell electrons in atoms. Still, my statement of scientific realism is intended to capture the possibility of both *optimistic* and *pessimistic* versions. An optimistic scientific realist thinks we can be confident that science is succeeding in uncovering the basic structure of the world and how it works. The pessimistic option is more cautious, even slightly skeptical. A pessimistic scientific realist might be someone who thinks that it is very hard for our feeble minds to get to the right theories, that evidence is often misleading, and that we tend to get too confident too quickly.

So there is a range of possible attitudes within scientific realism toward our chances of really understanding how the world works. Although there is a range, there is also a limit. My statement of scientific realism says that giving us accurate representations of the world is a *reasonable* aim of science. If someone thought it was just about *impossible* for us to get to the right theories, then it is hard to see how it could be a reasonable aim of science to *try* to do so. So there is a limit to the pessimism that is compatible with scientific realism as I understand it; *extreme* pessimism is not compatible. I think of Popper as someone who is getting close to this limit but who does not actually reach it.

Although Kuhn's most famous discussions of realism are his notorious claims about how the world changes when paradigms change, at other times he seems more like a pessimistic scientific realist. These are passages where Kuhn seems to think that the world is just so *complicated* that our theories will always run into trouble in the end—and this is a fact about the world that is independent of paradigms. We try to "force" nature into "boxes," but nature resists. All paradigms are doomed to fail eventually. This skeptical realist view is more coherent and more interesting than Kuhn's "changing worlds" position.

Much of the recent philosophical debate under the heading "scientific realism" has really been discussion of whether we should be optimistic or pessimistic about the aspirations of science to represent the world accurately (Psillos 1999). Some hold that fundamental ideas have changed so often within science—especially within physics—that we should always expect our current views to turn out to be wrong. Sometimes this argument is called the "pessimistic meta-induction." The prefix "meta" is misleading here, because the argument is not an induction about inductions; it's more like an induction about explanatory inferences. So let's call it "the pessimistic induction from the history of science." The pessimists give long lists of previously posited theoretical entities like phlogiston and caloric that we now think do not exist (Laudan 1981). Optimists reply with long lists of theoretical entities that once were questionable but which we now think definitely do exist—like atoms, germs, and genes.

These debates only have the ability to threaten scientific realism of the kind defended here it if they threaten to establish *extreme* pessimism. They do not support extreme pessimism. But the debates are interesting in their own right. What level of confidence should we have in our current theories, given the dramatic history of change in science? We should not think that this question is one to be settled *solely* by the historical track record. We might have reason to believe that our methods of hypothesizing and testing theories have improved over the years. But history will certainly give us interesting data on the question.

We might find good reason to have different levels of confidence, and also different *kinds* of confidence, in different domains of science. Ernan McMullin (1984) has rightly urged that we not think of the parts of physics that deal with the ultimate structure of reality as a model for all of science. Basic physics is where we deal with the most inaccessible entities, those furthest from the domain our minds are adapted to dealing with. In basic physics we often find ourselves with powerful mathematical formalisms that are hard to interpret. These facts give us grounds for caution. And where we are optimistic, we might have grounds for optimism about some features of our theories and not others. McMullin and also John Worrall (1989) have developed versions of the idea that the confidence we should have about basic physics is confidence that low-level *structural* features of the world have been captured reliably by our models and equations. That is a special *kind* of confidence.

All those factors that are relevant in the case of fundamental physics *do not apply* in the case of molecular biology. There we deal with entities that are far from the lowest levels, entities that we have a variety of kinds of access to. We do not find ourselves with powerful mathematical formalisms that are hard to interpret. The history of this field also supports a view holding that we are steadily accumulating knowledge of how biological molecules work and how they operate in the processes of life. So trying to work out the right attitude to have toward molecular biology is not the same as trying to work out the right attitude toward theoretical physics.

Realists sometimes claim that there is a general argument from the *success* of scientific theories to their truth. It is sometimes claimed that realism is the only philosophy of science that does not make the success of science into a miracle (Smart 1968; Putnam 1978). This line of argument has been unimpressive as a defense of realism. The real world will definitely have *some* role in affecting the success and failure of theories. Theories will do well or badly *partly* because of their relations to the world in which they are used and investigated. But there are *many* kinds of ways in which the link between theory and reality can generate success, especially in the short or medium term. Accurate representation of the world is not the only way. Theories can contain errors that compensate for each other. And theories can be successful despite being very wrong about the *kinds of things* they posit, provided they have the right *structure* in crucial places. Here is a simple example used by Laudan: Sadi Carnot thought that heat was a fluid, but he worked out some of the basic ideas of thermodynamics accurately despite this. The flow of a fluid was similar enough to patterns in the transfer of kinetic energy between molecules for his mistake not to matter much. Realists need to give up the idea that success in science points directly or unambiguously toward the truth of theories.

I hope my reasons for setting things up in the way I have are becoming clear. Much of the literature has held that scientific realists must be optimistic about current theories and about the history of science. I resisted that formulation of the issue. There is no point in arguing too much about the term "scientific realism," but there are benefits from organizing the issues in the way I have. What I call scientific realism is a fairly definite yes-or-no choice. (*Fairly* definite; see section 12.7.) This is also a choice about fundamental philosophical issues. The question about the right level of optimism to have about well-established scientific theories is *not* a question that has a simple answer that can be easily summarized. There we need to distinguish between different scientific fields, different kinds of theories, different kinds of success, and different kinds of optimism. In many cases we surely have good reason to be optimistic, but simple slogans should not be trusted.

One more comment on my formulation of scientific realism is needed. I said that science aims to give us "accurate descriptions and other representations of what reality is like." This is meant to be very broad, because there are lots of different kinds of representation used by different sciences. Some philosophers think that the main goal for a realist is *truth;* a good theory is a true theory. So they might want to formulate realism by saying that science aims to give us true theories. But the concepts of truth and falsity are only easy to apply in cases where a representation is in the form of language. In addition to linguistic representations, science often uses mathematical models, and other kinds of models, to describe phenomena. A scientific claim might also be expressed using a diagram. So I use the term "accurate representation" in a broad way to include true linguistic descriptions, pictures and diagrams that resemble reality in the way they are supposed to, models that have the right structural similarity to aspects of the world, and so on. I will return to these issues in the final section of this chapter.

12.4 Challenges from Traditional Empiricism

Scientific realism is now a popular position, but it has faced constant criticisms and challenges. Many of the most influential philosophers have thought that there is at least *something* wrong with scientific realism of the kind described in the previous section. Let's do a quick survey of the philosophers discussed so far in this book. Logical positivism was mostly opposed to scientific realism. Kuhn was vague and not always consistent, but he mostly opposed it. Many sociologists of science have certainly opposed it, including Latour. Goodman, the inventor of the "new riddle of induction," was opposed to it. Van Fraassen, who influenced my statement of what scientific realism is, rejects the view. So does Laudan. Feyerabend is hard to assess. Popper is in favor of scientific realism. Many of the naturalists discussed in the two previous chapters are scientific realists (including Fodor, Hull, and Kitcher), but not all are.

The critics listed above do not agree on *what* is wrong with scientific realism. I will divide the various forms of opposition into three broad families. Critics of realism differ among themselves just as much as they differ from the realists.

First, scientific realism has often been challenged by traditional forms of empiricism. In this book I will defend both scientific realism *and* a kind of empiricism, but this is not always an easy alliance. Indeed, one side of the debate about realism is often referred to as a debate *between* realism and empiricism.

Traditional empiricists tend to worry about both common sense and scientific realism, and they often worry for reasons having to do with knowledge. If there was a real world existing beyond our thoughts and sensations, how could we ever know anything about it? Empiricists believe that our senses provide us with our only source of factual knowledge. Many empiricists have thought that sensory evidence is not good enough for us to regard ourselves as having access to a "real world" of the kind the realist is committed to. And it seems strange (though not absurd, I think) to be in a position where you simultaneously say that a real world exists and also say we can never have any knowledge about it whatsoever.

The logical positivists recast these issues in terms of their theory of language. In the heyday of logical positivism, traditional philosophical questions about the "reality of the external world" were regarded as meaningless and empty. So the logical positivist attitude to most discussions of the "relation between science and reality" is that no side of the debate is saying anything meaningful and the whole discussion is a waste of time.

Some versions of logical positivism were also committed to the "phenomenalist" idea that all meaningful sentences can be translated into sentences that refer only to sensations. If phenomenalism is true, then when we *seem* to make claims about real external objects, all we are talking about are patterns in our sensations. Some more holistic empiricist views about language, of the kind associated with logical empiricism, have the same consequence. Even if translations are not possible, the nature of language prevents us from hoping to describe the structure of a world beyond our senses. Language and thought just cannot "reach" that far. I believe that a lot of twentieth-century empiricism held onto a version of this view (though some commentators on this book have objected to this claim).

In recent years the tension between realism and empiricism has often been debated under the topic of the "underdetermination of theory by evidence." Empiricists argue that there will always be a range of alternative theories compatible with all our actual evidence, and maybe a range of alternative theories compatible with all our possible evidence. So we never have good empirical grounds for choosing one of these theories over others and regarding it as representing how the world really is. This takes us back to the discussion in the previous section about the right level of optimism we should have about our scientific theories. I expressed scientific realism in a way compatible with a fair degree of pessimism, but the problem of underdetermination is important in its own right (see also sections 15.2 and 15.3).

12.5 Metaphysical Constructivism

I use the term "metaphysical constructivism" for a family of views including those of Kuhn and Latour. These views hold that, in some sense, we have to regard the world as *created* or *constructed* by scientific theorizing. Kuhn expressed this claim by saying that when paradigms change, the world changes too. Latour expresses the view by saying that nature (the real world) is the *product* of the decisions made by scientists in the settlement of controversies. Nelson Goodman is another example; he argues that when we invent new languages and theories, we create new "worlds" as well (1978). For a metaphysical constructivist, it is not even *possible* for a scientific theory to describe the world as it exists independent of thought, because reality itself is dependent on what people say and think.

These views are always hard to interpret, because they look so strange when interpreted literally. How could we possibly *make* the world just by making up a new theory? Maybe Kuhn, Latour, and Goodman are just using a metaphor of some kind? Perhaps. Kuhn sometimes expressed a different view on the question, a kind of skeptical realism, and he struggled to make his position clear. But when writers such as Goodman have been asked about this, they have generally insisted that their claims are *not* just metaphorical (Goodman 1996, 145). They think there is something quite wrong with scientific realism of the kind I described in section 12.3. They accept that it's hard to describe a good alternative, but they think we should use the concept of "construction," or something like it, to express the relationship between theories and reality.

Some of these ideas can be seen as modified versions of the view of Immanuel Kant ([1781] 1998). Kant distinguished the "noumenal" world from the "phenomenal" world. The noumenal world is the world as it is in itself. This is a world we are bound to believe in, but which we can never know anything about. The phenomenal world is the world as it appears to us. The phenomenal world is knowable, but it is partly our creation. It does not exist independently of the structure of our minds.

This kind of picture has often seemed appealing to philosophers who want to deny scientific realism but do so in a moderate way. Hoyningen-Huene (1993) has argued that we should interpret Kuhn's views as similar to Kant's. In Michael Devitt's analysis of the realism debates (1997), a wide range of philosophers are seen as either deliberately or inadvertently following the Kantian pattern. According to Devitt, constructivist antirealism works by combining the Kantian picture with a kind of relativism, with the idea that different people or communities create different "phenomenal worlds" via the imposition of their different concepts on experience. This relativist idea was not part of Kant's original view; for Kant all humans apply the same basic conceptual framework and have no choice in the matter.

The Kantian picture is sometimes seen as a way of holding onto the idea that there is a real world *constraining* what we believe but doing so in a way that does not permit our knowing or representing this world. This move is often tempting, but the resulting views are unhelpful. Understanding our access to reality is difficult, but adding an extra layer called "the phenomenal world" in between us and the real world achieves nothing.

The term "social constructivism" is often used for roughly the same kind of view that I am calling metaphysical constructivism. But "social constructivism" is also used for more moderate ideas as well. If someone argues that we make or construct our *theories,* or our classifications of objects, that claim is not opposed to scientific realism. We do indeed "construct" our ideas and classifications. Nature does not hand them to us on a platter. But a scientific realist insists that beyond ideas and theories there is also the rest of reality.

In fields like sociology of science, as we saw in chapter 8, there is an unfortunate tradition of not explicitly distinguishing between the construction of ideas and the construction of reality. What is it about these fields that has encouraged such strange-sounding formulations of ideas? There are various reasons, but I will venture some meta-sociology here—sociology of the sociology of science. A lot of work in these fields has been organized around the desire to oppose a particular *Bad View* that is seen as completely wrong. The Bad View holds that reality determines thought by stamping itself on the passive mind; reality acts on scientific belief with "unmediated compulsory force" (Shapin 1982, 163). That picture is to be avoided at all costs; it is often seen as not only false but even *politically* harmful, because it suggests a passive, inactive view of human thought. Many traditional philosophical theories are interpreted as implicitly committed to this Bad View. This is one source for descriptions of logical positivism as reactionary, helpful to oppressors, and so on.

What results from this is a tendency for people to go *as far as possible* away from the Bad View. This encourages people to assert simple *reversals* of the Bad View's relationship between mind and world. Thus we reach the idea that theories construct reality.

Some explicitly embrace the idea of an "inversion" of the traditional picture (Woolgar 1988, 65), while others leave things more ambiguous. But there is little pressure within the field to discourage people from going too far in these statements. (Bloor 1999 is an interesting exception.) Indeed, those who express more moderate denials of the Bad View leave themselves vulnerable to criticism from within the field. The result is a literature in which one error—the view that reality stamps itself on the passive mind—is exchanged for another error, the view that thought or theory constructs reality.

12.6 Van Fraassen's View

The last form of opposition to scientific realism that I will discuss is a more moderate and careful form; this is the position of Bas van Fraassen (1980). Van Fraassen's ideas lie within the empiricist tradition, but they are not based on a linguistic or psychological theory. Instead, van Fraassen confronts realism on the proper aims of science. So his antirealism is a direct denial of the kind of scientific realism defended in this chapter. This is no accident, since my formulation of scientific realism was influenced by his.

In discussions of realism, the term "instrumentalism" is used to refer to a variety of antirealist views. Sometimes it is used for traditional empiricist positions of the kind discussed earlier. But sometimes it is used in a different way, which I think is more appropriate. According to instrumentalism in this sense, we should think of scientific theories as devices for helping us deal with experience. Rather than saying that describing the real world is impossible, an instrumentalist will urge us *not to worry* about whether a theory is a true description of the world, or whether electrons "really, really exist." If a theory enables us to make good predictions, what more can we ask? If we have a theory that gives us the right answers with respect to what we can observe, we might occasionally find ourselves wondering if these right answers result from some deeper "match" between the theory and the world. But we can never expect to know the answer to this question, so what relevance does it have to science? Quite a few scientists have expressed instrumentalist views, especially in physics. The idea that we should ignore questions about the "real reality" of theoretical entities because these questions have no practical relevance is also linked to one strand of the *pragmatist* tradition in philosophy (Rorty 1982).

A detailed version of this kind of position has been worked out by van Fraassen (1980). Van Fraassen does not use the term "instrumentalist" to describe his view; he calls it "constructive empiricism." The term "constructive" is used by so many people that it often seems to have no meaning at all, so I have reserved it for the views discussed in section 12.5. I see van Fraassen's view as a version of the instrumentalist approach, but it does not matter much what we call it.

Van Fraassen suggests that all we should ask of theories is that they accurately describe the observable parts of the world. Theories that do this are "empirically adequate." An empirically adequate theory *might* also describe the hidden structure of reality, but whether or not it does so is of no interest to science. For van Fraassen, when a theory passes a lot of tests and becomes well established, the right attitude to have toward the theory is to "accept" it, in a special sense. To accept a theory is to (1) believe (provisionally) that the theory is empirically adequate, and to (2) use the concepts the theory provides when thinking about further problems and when trying to extend and refine the theory.

Regarding point 1, for a theory to be empirically adequate, it must describe *all* the observable phenomena that come within its domain, including those we have not yet investigated. Some of the familiar problems of induction and confirmation appear here. Regarding point 2, van Fraassen wants to recognize that scientists do come to "live inside" their theories; they make use of the theory's picture of the world when exploring new phenomena. Some versions of instrumentalism struggle to make sense of this fact. But van Fraassen says a scientist can "live inside" a theory while remaining agnostic about whether the theory is true.

How can we decide between van Fraassen's view and the version of scientific realism that I outlined earlier?

First we need to be sure that the two positions conflict. I said that *one* aim of science is to give us accurate representations of the world, including the unobservable parts. Van Fraassen says "science aims to give us theories which are empirically adequate" (1980, 12). So far, our views seem compatible. In some cases science could aim only at empirical adequacy, but in other cases it could aim at representing the hidden structure of the world as well.

And this is the *right* attitude for a realist to have. For various reasons and in various situations, it might make sense for a scientist to be cautious, or unconcerned, about the application of a theory to the unobserved structure of the world, even when he or she is becoming confident about empirical adequacy.

So van Fraassen has described an attitude that scientists can reasonably have toward *some* theories in *some* circumstances. But van Fraassen thinks that science *should aim at no more* than empirical adequacy.

As many have argued, one place where van Fraassen's view runs into trouble is the distinction between observable and unobservable parts of the world. Realists have argued that there is a continuum, rather than a sharp boundary, between the observable and the unobservable (Maxwell 1962). Some things can be observed with the naked eye, like trees. Other things, like the smallest subatomic particles, are unobservable and can only have their presence inferred from their effects on the behavior of observable things. But between the clear cases we have lots of unclear ones. Is it observation if you use a telescope? How about a light microscope? An X-ray machine? An MRI scan? An electron microscope? The realist thinks that the distinction between the observable and unobservable is vague, and not of the right kind to support general conclusions about what science aims to represent.

Van Fraassen accepts that the distinction between the observable and the unobservable is vague, and he accepts that there is nothing "unreal" about the unobservable. He also accepts that we learn about the boundary from science itself. Still, he argues, science is only concerned with empirical adequacy—making true claims about the observable part of the world. But this view cannot be defended. Van Fraassen is saying it's *never* reasonable for science to aim at describing the structure of the world beyond *this particular* boundary. Suppose we describe a slightly different boundary, based on a concept a bit broader than "observation." Let's say that something is *detectable* if either it is observable or its presence can be very reliably inferred from what is observable. As with van Fraassen's concept of observability, science itself tells us which things are detectable. In this sense, the chemical structures of various important molecules like sugars and

DNA are detectable although not observable. So why shouldn't science aim at giving us accurate representations of the detectable features of the world as well as the observable features? Why shouldn't science aim to tell us what the molecular structure of complex sugars is like?

Perhaps our beliefs about the detectable structures are not as reliable as our beliefs about the observable structures. If so, we need to be more cautious when we take theories to be telling us what the detectable structure of the world is like. But that is no problem; we often need to be cautious.

What is so special about the "detectable"? Nothing, of course. We could define an even broader category of objects and structures, which includes the detectable things plus those that can have their presence inferred from observations with *moderate* reliability. Why should science stop before trying to work out what lies beyond this boundary? We might need to be even *more* careful with our beliefs about those features of the world, but that is no problem.

You can see how the argument is going. From the realist point of view, there is no boundary that marks the distinction between features of the world that science can reasonably aim to tell us about and features that science cannot reasonably aim to tell us about. As we learn about the world, we also learn more and more about which parts of the world we can expect to have reliable information about. And there is no reason why science should not try to describe all the aspects of the world that we can hope to gain reliable information about. As we move from one area to another, we must often adjust our level of confidence. Sometimes, especially in areas such as theoretical physics, which are fraught with strange puzzles, we might have reason to adopt something like van Fraassen's attitude, at least temporarily. But it is a mistake to think that empirical adequacy of van Fraassen's kind is *the* aim of science.

12.7 Representation, Models, and Truth (Optional Section)

I will finish the chapter with further discussion of an issue introduced in section 12.3. I formulated scientific realism by saying that science tries to give us "accurate representations" of the world. Most discussion of this topic in twentieth-century philosophy treated theories as *linguistic* entities, as collections of sentences. So when people tried to work out what sorts of relationships theories have with reality, they drew on concepts from the philosophy of language. In particular, the concepts of *truth* and *reference* were emphasized. A good scientific theory is a true theory; how can we determine which theories are true? Electrons exist if the word "electron" refers to them; how do we decide whether a term in a scientific theory refers to anything? A range of problems came to be addressed via the concepts of truth and reference.

This emphasis might be a bad idea. There are several issues to consider. One has to do with the "representational vehicles," or representational media, used by science. Science does express hypotheses about the world using sentences in language, either ordinary language or technical extensions of ordinary language. But in other cases, science uses representational vehicles of a different kind. Many hypotheses in science are expressed using *models*. Consider the case of mathematical models. These are abstract mathematical structures that are supposed to represent key features of real systems in the world. But in thinking about how a mathematical model might succeed in representing the world, the linguistic concepts of truth, falsity, reference, and so forth do not seem to be useful. Models have a different *kind* of representational relationship with the world from that found in language. A good model is one that has some kind of *similarity* relationship, probably of an abstract kind, with the system that the model is "targeted" at (Giere 1988). It is hard to work out the details of this idea.

The role of models in science did become an important topic in late-twentieth-century philosophy (Suppe 1977). Some argued that we should use the idea of a model to give a different description of how *all* theories work in science. But it is a mistake to think that all of science uses the same "vehicles" to represent the world. We should not replace a language-based analysis of all science with a model-based analysis. What we find in science is a range of different representational vehicles.

Consider Darwin's *Origin of Species*. Darwin's book contained a set of hypotheses about the world, supported with elaborate arguments, expressed using rather ordinary language. But not all science is like this. Even the topics that Darwin was addressing are now treated differently. Recent discussions of how natural selection changes biological populations tend to be expressed in the form of mathematical models. These models are written down, of course. They are formulated using mathematical symbolism, and they have to be supplemented with a commentary telling us (for example) which phenomena in the real world are being represented by the model. But we should not expect an analysis of how mathematical models relate to the world to use the same concepts as an analysis of how hypotheses expressed in ordinary language relate to the world.

Not all models in science are mathematical. More generally, we might think of a model as a structure that is intended to represent another structure by virtue of an abstract similarity relationship between them. Sometimes the aim might be to understand the unfamiliar by modeling it on the familiar (as in Bohr's early "solar system" model of the atom). But this is not always what is going on. Abstract mathematical models might be thought of as attempts to use a general-purpose and precise framework to represent *dependence relationships* that might exist between the parts of real systems. A mathematical model will treat one variable as a function of others, which in turn are functions of others, and so on. In this way, a complicated network of dependence structures can be represented. And then, via a commentary, the dependence structure in the model can be treated as representing the dependence structure that might exist in a real system.

Models, whether mathematical or not, have a kind of flexibility that is important in scientific work. A variety of people can use the same model while interpreting it differently. One person might use the model as a predictive device, something that gives an output when you plug in specific inputs, without caring how the inner workings of the model relate to the real world. Another person might treat the same model as a highly detailed picture of the dependence structure inside the real system being studied. And there is a range of possible attitudes between these two extremes; another person might treat the model as representing *some* features, but only a few, of what is going on in the real system.

The difference between models and linguistically expressed theories may be important in understanding progress in science. Many old scientific theories, now superseded, can look like failures when we ask whether much of the theory was *true,* and whether the terms in the theory *referred* to anything. But sometimes, if we recast the old theory as a model, we find that the model had some of the right *structure,* from the point of view of our current theories. Worrall (1989) uses the case of various "ether" theories from nineteenth-century physics; they had good structural features even though the ether does not exist.

In criticizing the emphasis on truth and reference in philosophy of science, I have stressed the role of representational vehicles that require a different kind of analysis. Some would add that even when we are dealing with language, the concepts of truth and reference might be bad ones to use.

Some philosophers think that to call a theory true is to assert that it has a special connection to the world. Traditionally, this has been described as a *correspondence* relationship. That term can be misleading, as it suggests a kind of "picturing," which is not what modern theories of truth propose. But this first option holds that there is some kind of special and valuable relationship between true theories and the world. If this is so, we can use the concept of truth when analyzing scientific language and its relations to reality. Others argue that the concept of truth is *not* suitable for this kind of use. The word "true" is one that we use to signal our agreement or disagreement with others, not to describe real connections between language and the world (Horwich 1990). In sociology of science, Bloor (1999) has defended a position of this kind.

In this chapter I have been cautious about truth. I used a broad concept of "accurate representation" to describe a goal that science has for its theories. Some argue that even the idea of *representation* as a genuine relationship between symbols and the world is mistaken, whether the symbols are in language, models, thought, or whatever. That will sound like a radical position, and so it is. (This is one claim made by postmodernists, for example.) But it is hard to work out which theories about symbols retain the familiar idea of representation, and which do not.

Further Reading

Key works in the resurgence of scientific realism include Jack Smart's *Philosophy and Scientific Realism* (1963) and various papers collected in Hilary Putnam's *Mind, Language, and Reality* (1975). See also Maxwell 1962.

Leplin, *Scientific Realism* (1984), is a very good collection on the problem. Boyd's paper in that book is a useful survey of the options, with key differences from the one given here. Boyd also gives an influential defense of scientific realism. Devitt, *Realism and Truth* (1997), defends both common-sense and scientific realism. Psillos 1999 is a very detailed treatment of the debate.

For further discussion of the relations between realism and success, see Stanford 2000. On the problems raised by quantum physics, see Albert 1992. For a more detailed discussion of how avoidance of the "Bad View" has shaped sociology of science, see Godfrey-Smith 1996, chapter 5.

Churchland and Hooker, *Images of Science* (1985), is a good collection on van Fraassen.

Kitcher (1978) battles with the problems of meaning and reference for scientific language and their consequences for realism. See also Bishop and Stich 1998 on this problem. Lynch 2001 is a recent collection on the problem of truth.

There is a large literature on the role of models in science (Suppe 1977). Confusion sometimes arises because the usual sense of the word "model" in philosophy is different from that found in science itself (see the glossary). So different people wanting to "analyze science in terms of models" often have very different tasks in mind (Downes 1992). One useful and interesting treatment of the issue is in Giere's *Explaining Science* (1988, chapter 3). Hesse 1966 is a famous early discussion, focused, however, on yet another sense of "model."

Fine 1984 and Hacking 1983 are influential works on scientific realism that defend rather different views from those discussed here.

13

Explanation

13.1 Knowing Why

What does science do for us? In chapter 12 I argued for a version of scientific realism, according to which one aim of science is describing the real structure of the world. Science aims to tell us, and often succeeds in telling us, what the world is like. But it is also common to think that science tells us *why* things happen; we learn from science not just what goes on but why it does. Science apparently seeks to *explain* as well as describe. So we seem to face a new question. What *is* it for a scientific theory to explain something? In what sense does science give us an *understanding* of phenomena, as opposed to mere descriptions of what there is and what happens?

The idea that science aims at explanations of *why* things happen has sometimes aroused suspicion in philosophers, and it has also done so in scientists themselves. Such distrust is reasonably common within strong empiricist views. Empiricists have often seen science, most fundamentally, as a system of rules for predicting experience. When explanation is put forward as an *extra* goal for scientific theories, empiricists get nervous.

There is a complicated relationship between this problem of explanation and the problem of analyzing confirmation and evidence (chapters 3, 14). The hope has often been to treat these problems separately. Understanding evidence is problem 1; this is the problem of analyzing what it is to have evidence to believe that a scientific theory is true. Understanding explanation is problem 2; here we assume that we have already chosen our scientific theories, at least for now. We want to work out how our theories provide explanations. In principle, we can make a distinction of this kind. But there is a close connection between the issues. The solution to problem 2 may affect how we solve problem 1. Theories are often preferred by scientists because they seem to yield good explanations of puzzling phenomena. In chapter 3, *explanatory inference* was defined as inference from a set of data to a hypothesis about a structure or process that would explain the data. This seems much more common in science than the traditional philosophical idea of inductive inference (inference from particular cases to generalizations). This suggests that there is a close relation between the problem of analyzing explanation and the problem of analyzing evidence.

There is a very large literature on explanation, but these issues will get a whirlwind treatment in this book. One reason for this is that I think the philosophy of science has approached the problem of explanation in a mistaken way. To some extent, that is true of many topics in this book; there have been plenty of wrong turns in the philosophy of science. But in the case of explanation, I think the error has been fairly clear; I will describe it in section 13.3. So some of the views presented in this chapter are rather unorthodox.

13.2 The Rise and Fall of the Covering Law Theory of Explanation

Empiricist philosophers, I said above, have sometimes been distrustful of the idea that science explains things. Logical positivism is an example. The idea of explanation was sometimes associated by the positivists with the idea of achieving deep metaphysical insight into the world—an idea they would have nothing to do with. But the logical positivists and logical empiricists did make peace with the idea that science explains. They did this by construing "explanation" in a low-key way that fitted into their empiricist picture.

The result was the *covering law* theory of explanation. This was the dominant philosophical theory about scientific explanation for a good part of the twentieth century. The view is now dead, but its rise and fall are instructive.

The covering law theory of explanation was first developed in detail by Carl Hempel and Paul Oppenheim in a paper (1948) that became a centerpiece of logical empiricist philosophy. Let us begin with some terminology. In talking about how explanation works, the *explanandum* is whatever is being explained. The *explanans* is the thing that is doing the explaining. If we ask "why X?" then X is the explanandum. If we answer "because Y," then Y is the explanans.

The basic ideas of the covering law theory are simple. Most fundamentally, to explain something is to *show how to derive it in a logical argument.* The explanandum will be the conclusion of the argument, and the premises are the explanans. A good explanation must first of all be a good logical argument, but in addition, the premises must contain at least one statement of a *law of nature*. The law must make a real contribution to the argument; it cannot be something merely tacked on. (Of course, for an explanation to be a good one in the fullest sense, the premises must also be *true*. But the first task here is to describe what sort of statements *would* give us a good explanation of a phenomenon, if the statements were true.)

Some explanations (both in science and in everyday life) are of particular events, while others are directed at general phenomena or regularities. For example, we might try to explain the particular fact that the U.S. stock market crashed in 1929, in terms of economic laws operating against the background conditions of the day. And we can also explain patterns; Newton is often seen as giving an explanation of Kepler's laws of planetary motion in terms of more basic laws of mechanics in combination with assumptions about the layout of the solar system. In both cases, the covering law theory sees these explanations as expressible in terms of arguments from premises to conclusions. Some of the arguments that express explanations will be deductively valid, but this is not required in all cases. The covering law theory was intended to allow that some good explanations could be expressed as nondeductive arguments ("inductive" arguments, in the logical empiricists' broad sense of the term). If we can take a particular phenomenon and embed it within an argument in which the premises include a law and bestow high probability on the conclusion, this yields a good explanation of the phenomenon.

There were many problems of detail encountered in attempts to formulate the covering law theory precisely (Salmon 1989). The problems were more difficult in the case of nondeductive arguments, and also in the case of explaining patterns rather than particular events. I won't worry about the technicalities here. The basic idea of the covering law theory is simple and clear: to explain something is to show how to derive it in a logical argument of a kind that makes use of a law in the premises. To explain something is to *show that it is to be expected*, to show that it is *not surprising*, given our knowledge of the laws of nature.

For the covering law theory, there is not much difference between explanation and prediction. To predict something, we put together an argument and try to show that it is to be expected, though we don't know for sure yet whether it is going to happen. When we explain something, we know that it has happened already, and we show that it *could* have been predicted, using an argument containing a law. You might be wondering at this point what a "law of nature" is supposed to be. This was a troubling topic for logical empiricism and has continued to be troubling for everyone else. But a "law of nature" was not supposed to be something very grandiose. It was supposed to be a kind of basic regularity, a basic pattern, in the flow of events. (I return to this question in section 13.4.)

Though I use the phrase "covering law theory" here, another name for the theory is the "D-N theory," or "D-N model," of explanation. "D-N" stands for "deductive-nomological," where the word "nomological" is from the Greek word for law, *nomos*. The term "D-N" can be confusing because, as I said, the argument in a good explanation need not be deductive. So "D-N" really only refers to some covering law explanations, the deductive ones.

That concludes my sketch of the covering law theory. I now move on to what is wrong with it. This is a case where we have something close to a knockdown argument. Although there are many famous problems with the covering law theory, the most convincing problem is usually called the *asymmetry problem*. And the most famous illustration of the asymmetry problem is the case of the flagpole and the shadow.

Suppose we have a flagpole casting a shadow on a sunny day. Someone asks: why is the shadow *X* meters long? According to the covering law theory, we can give a good explanation of the shadow by deducing the length of the shadow from the height of the flagpole, the position of the sun, the laws of optics, and some basic trigonometry. We can show why that length of shadow was to be expected, given the laws and the circumstances. The argument can even be made deductively valid. So far, so good. The problem is that we can run just as good an argument in another direction. Just as we can deduce the length of the shadow from the height of the pole (plus optics and trigonometry), *we can deduce the height of the pole from the length of the shadow* (and the same laws). An equally good *argument*, logically speaking, can be run in both directions; either can give information about the other. But it seems that we cannot run an equally good *explanation* in both directions, though the covering law theory says we can. It is fine to explain the length of the shadow in terms of the flagpole and the sun, but it is not fine to explain the length of the flagpole in terms of the shadow and the sun. (At least, it is not fine unless this is a very unusual flagpole—perhaps one that is designed to regulate its own length to maintain a particular shadow.)

What we find here is that explanations have a kind of directionality. Some arguments (though not all) can be reversed and remain good as arguments. But explanations cannot be reversed in this way (except in some special cases). So not all good arguments that contain laws are good explanations. This objection to the covering law theory was famously given (using a slightly different example) by Sylvain Bromberger (1966).

Once this point is seen, it becomes obvious and devastating. The covering law theory sees explanation as very similar to prediction; the only difference is what you know and what you don't know. But this is a mistake. Consider the concept of a *symptom*. Symptoms can be used to predict, but they cannot be used to explain. Yet a symptom can often be used in a good logical argument, along with a law, to show that something is to be expected.

If you know that only disease D produces symptom S, then you can make inferences from S to D. You might in some cases be able to make predictions from D to S, too. But you cannot *explain* a disease in terms of a symptom. Explanation only runs one way, from D to S, no matter how many different kinds of inferences can be made in other directions. And, further, it seems that good explanations of S can be given in terms of D even if S is *not* a very reliable symptom of D, even if S is not always to be expected when someone has D. This is a separate problem for the covering law theory, often discussed using the example of some unreliable but unpleasant symptoms of syphilis.

In some of these cases, the covering law theory can engage in fancy footwork to evade the problem. But other cases, including the original flagpole case, seem immune to footwork. Hempel's own attitude to the issue was puzzling. He actually anticipated the problem, but dismissed it (Hempel 1965, 352–54). His strategy was to accept that if his theory allowed explanations to run in two directions in cases where it seems that explanation only runs in one direction, then both directions must really be OK. In some actual scientific cases this reply seems reasonable; there are cases in physics where it *is* hard to tell which direction(s) the explanation(s) are running in. But in other cases the direction seems completely clear. In the case of the flagpole and the shadow, this reply seems hopeless.

There are other good arguments against the covering law theory (Salmon 1989), but the asymmetry problem is the killer. It also seems to be pointing us toward a better account of explanation.

13.3 Causation, Unification, and More

What is it about the flagpole's height that makes it a good explanation for the length of the shadow, and not vice versa? The answer seems straightforward. The shadow is *caused* by the interaction between sunlight and the flagpole. That is the direction of causation in this case, and that is the direction of explanation also. So we seem to get an immediate suggestion from the flagpole case for how to build a better theory: to explain something is to describe what caused it. Why did the dinosaurs become extinct 65 million years ago? Here again, our request for an explanation seems to amount to a request for information about what caused the extinction.

Although that conclusion seems compelling, it has not been universally accepted, and it raises many further problems. The biggest question raised at this point is, What *is* causation? We confidently used the idea of causation to resolve the flagpole case, but the whole idea of causation and causal connection is extremely controversial in philosophy. For many philosophers, causation is a suspicious metaphysical concept that we do best to avoid when trying to understand science. This suspicion is, again, common within the empiricist tradition. It derives from the work of Hume. The suspicion is directed especially at the idea of causation as a sort of hidden connection between things, unobservable but essential to the operation of the universe. Empiricists have often tried to understand science without supposing that science concerns itself with alleged hidden connections of that kind. The rise of scientific realism in the latter part of the twentieth century led to some easing of this anxiety. But many philosophers would be pleased to see an adequate account of science that did not get entangled with issues about causation.

Despite this unease, toward the end of the twentieth century, the main proposal about explanation being discussed, in different ways, was the idea that explaining something is giving information about how it was caused. Some sophisticated analyses were developed that sought to use probability theory to clarify this basic idea (Salmon 1984; Suppes 1984). It might seem initially that this view of explanation is most directly applied to explanations of particular events (like the extinction of the dinosaurs), but it can also be applied to the explanation of patterns. We can ask, Why does inbreeding produce an increase in birth defects? The explanation will describe a general kind of causal process that is involved in producing the phenomenon (a process involving an increased chance that two copies of a recessive gene will be brought together in a single individual).

Claiming that causation is the key to explanation does not settle all the issues about explanation. We need to know what *kind* of information about causes is needed for a good explanation. One way to think of the situation is to imagine an idealized "complete" explanation that contains *everything* in the causal history of the event to be explained, specified in total detail (Railton 1981). No one ever wants to be told the complete explanation for a phenomenon, and we never know these complete explanations. Instead, in any context of discussion in which a request for an explanation is made, some *piece* or *pieces* of the complete explanation will be relevant. We are often able to know, and describe, these relevant pieces of a total causal structure. To give a good explanation in actual practice, all that is required is a description of these relevant pieces of the whole.

One main alternative to the analysis of explanation in terms of causation was developed in the years after the demise of the covering law theory. This was the idea that explanation should be analyzed in terms of *unification.* This idea was developed in detail by Michael Friedman (1974) and Philip Kitcher (1981, 1989). But as Kitcher also emphasized, the idea was actually present all along in logical empiricism. The idea that explanation is unification was a sort of "unofficial" theory of explanation within much logical empiricism, in contrast to the "official" covering law theory (see, for example, Feigl 1943). This unofficial theory is a good deal better than the official theory. Often the two approaches were mixed in together; to show the connection between particular events and a general law is, after all, to achieve a kind of unification. Why not develop a theory of unification in science that is not tied to the idea of deriving phenomena from laws?

So the *unificationist* theory holds that explanation in science is a matter of connecting a diverse set of facts by subsuming them under a set of basic patterns and principles. Science constantly strives to reduce the number of things that we must accept as fundamental. We try to develop general *explanatory schemata* (explanatory schemes) that can be applied as widely as possible. This proposal certainly makes a lot of sense of how scientists operate. Indeed, it seems clear that what produces an "Aha!" reaction is often the realization that some odd-looking phenomenon is really a case of something more general. Kitcher also argues for this view with cases from the history of science. He argues that some very famous theories—Darwin's theory of evolution and Newton's later work on the nature of matter—were compelling to scientists in their early stages despite not making many specific new predictions, because they promised to *explain so much*. And this "explanatory promise" seems to have been the ability of those theories to *unify* a great range of phenomena with a few general principles.

In Kitcher's case, another reason for developing the unificationist theory was a distrust of the idea of causation. This led Kitcher to try, for some years, to develop a theory of explanation *entirely* in terms of unification. But what about the flagpole and the shadow, and the asymmetry in which can explain which? Kitcher argued that we do tend to describe this asymmetry in causal terms, but this causal talk is really a loose summary of more basic asymmetries that involve unification (Kitcher 1989).

So we have two main proposals to replace the covering law theory: the causal theory and the unification theory. These have often been treated as competitors: "Does causation win or does unification win?" But this is surely a mistake. We do not have to choose. Again, beware the dubious allure of simplicity in philosophical theories! Much of the time, to explain something is to describe the causal mechanisms behind it or the causal history leading up to it. That is true *much* of the time, but there is no need to hold that it is true *all* the time. In some cases there can be pretty clear explanatory relations between patterns or principles, even when causal language is hard to apply to the situation. Often this seems to involve unification. Nothing stops us from holding that a variety of different relations can be explanatory.

Recently, ideas similar to this have been emerging in the philosophy of science. Wesley Salmon was for many years one of the main partisans for the idea that causation is the key to explanation. But he eventually accepted that unification is also part of the story. Sometimes he seemed to think of causation and unification as two sides of the same explanatory coin, and some other times as alternative explanatory projects (1989, 1998). Kitcher, who tried for years to avoid using the idea of causation to analyze explanation, instead telling the whole story in terms of unification, has now decided that this was probably a mistake and the concept of causation is not so dubious after all (personal communication, 2002).

So what might be emerging is a kind of "pluralism" about explanation in the philosophy of science. This is a step in the right direction, but I suggest that the whole issue has been approached wrongly. (This is where I become unorthodox.) The most peculiar thing about the discussion of explanation by philosophers has been the assumption that explanation is the kind of thing that requires analysis in terms of a single special relation or a short list of special relations. It is a mistake to think there is one basic relation that is *the* explanatory relation (as in the covering law theory, the causal theory, and the unification theory), and it is also a mistake to think that there are some definite two or three such relations.

The alternative view is to recognize that the idea of explanation operates differently within different parts of science—and differently within the same part of science at different times. The word "explanation" is used in science for something that is sought, and sometimes achieved, by the development of theories, but what exactly is being sought is not constant in all of science. And we cannot get the right analysis by claiming that within all of science, a good explanation is something that satisfies *either* the causal test *or* the unification test (etc.). This familiar form of "pluralism" leaves out the way that different scientific fields will establish definite criteria for what will pass as a good explanation. The standards for a good explanation in field A need not suffice in field B. If an "ism" is required, the right analysis of explanation is a kind of *contextualism*—a view that treats the standards for good explanation as partially dependent on the scientific context.

Kuhn argued some years ago for a view of this kind (1977a). In a paper about the history of physics, he claimed that different theories (or paradigms) tend to bring with them their own standards for what counts as a good explanation. He argued, further, that standards about whether a relation counts as "causal" also depend on paradigms. The concepts of explanation and causation are, to some extent at least, internal to different scientific fields and historical periods.

In the case of causation, a philosopher might reply to Kuhn, with some justification, that just because different people have *thought* differently about what causation is does not mean that there *is* no fact of the matter. Fair enough (at least for now). But in the case of explanation, I think this reply has little force. If two scientific fields single out different relations and call them "explanation," there need be no factual error that one or the other is making.

To support this claim, Kuhn focused (as he did in *Structure*) on the case of Newton's theory of gravity. Does Newton's theory *explain* the falling of objects, given that Newton's treatment of gravity gave no intuitive mechanism but only a mathematical relationship? Some answered no, but over time it became part of Newtonianism that the right kind of mathematical law *does count* as an explanation. It is Kuhn's view that the idea of explanation will evolve as our ideas about science and about the universe change.

So although the covering law theory definitely fails as a general account of explanation in science, it would be a mistake to conclude that *no* explanations have the form described by the covering law theory. There are some explanations that are at least close to what Hempel had in mind. The mistake is to apply that model to all cases.

I suggest that Kuhn was right on this point. I add that this proposal need not lead to the radical idea that *anything* can count as an explanation. Scientific traditions will generally have good reasons for their treatment of the idea of explanation; views about explanation will depend on views about what the world contains, for example. Some possible concepts of explanation will embed factual errors. To use a simple example, if someone claims that good explanations are always based on a concept of God's will, but it turns out that there is no God, then that conception of explanation will be mistaken because of a factual error. Some philosophers might make the same argument about concepts of explanation that use the idea of causation—they might argue that the traditional idea of causal connection is a piece of mistaken metaphysics. But many possible treatments of the idea of explanation will not be ruled out by factual errors.

This is a case where it is important to pay attention to the actual use of the term "explain" in science. Here we find a lot of diversity. In some fields, there are technical senses of the word, even mathematical measurements of "the amount of variation explained" by a given factor. In other fields, nothing like a technical standard applies. The word "explain" also has an almost rhetorical use. Someone might say: "your theory does accommodate this result, but it does not really explain it." This might mean "your theory can only be used to derive this result in an unnatural-looking way." (Often, unification seems important in cases like this.) At other times the word "explain" is used in a much more low-key way in science. According to scientific realism, a lot of science is aimed at describing what is going on in the world; often this will be a matter of describing *how things work*. How does photosynthesis work? How does the replication of DNA work? Descriptions of phenomena of this kind will often be referred to as explanations, but this does not mean that something *extra* is going on, beyond the description of mechanisms and processes.

At this point I should compare my view with another unorthodox position in this area, that of van Fraassen (1980). He denies, as I do, that explanation is some single, special relation common to all of science. He has developed a "pragmatic" account of explanation, in which what counts as an explanation varies according to context. But this is very different from my view. Van Fraassen wants to deny that explanation is something "inside" science at all; he denies that scientific reasoning includes the assessment of the explanatory power of theories. Instead, explanation is something that people do when they take scientific theories and use them to answer questions that are external to scientific discussion itself. In contrast, the view I am defending is a view in which explanation is thoroughly internal to science, *but variously so*. Assessments of what explains what are a very important part of scientific reasoning, but different fields may use somewhat different concepts and standards of explanation.

Before leaving this topic, I should also add a comment about *explanatory inference*. Back in chapter 3, I used this term for inference from a set of data to a hypothesis about a structure or process that would explain the data. In chapter 14, when I look at more recent discussions of confirmation and evidence, I will return to this topic. The term "explanatory inference" suggests that there is one kind of relationship between data and hypothesis—*the* explanation relationship—that is involved in explanatory inference. Many philosophers would accept this. The term "inference to the best explanation" is, in fact, a more common name for what I call explanatory inference; that term suggests that there is some single measure of "explanatory goodness" involved. But I think this is the wrong way to think about scientific reasoning (and this is why I have avoided the term "inference to the best explanation"). I use "explanatory inference" in a broader way that does not suppose that there is a single measure of explanatory goodness involved, which applies to all of science. Rather, explanatory inference is a matter of devising and comparing hypotheses about hidden structures that might be responsible for data. "Explanation" is seen as something pretty diverse.

To sum up: the covering law theory is dead, as a general account of explanation in science. But we should not look for a new theory of some single relation or pattern that is involved in all scientific explanation. Very often, causation is involved. The same goes for unification and for deriving phenomena from laws. But different fields have different concepts and standards of explanation.

13.4 Laws and Causes (Optional Section)

This short and rather abstract section is a digression from the main themes of the book, a foray into an intersection between philosophy of science and the controversial field of *metaphysics*. The covering law theory of explanation made use of the idea of a law of nature. One of the theories that replaced it made use of the idea of causation. But what are laws of nature? What is causation?

In both cases, we have concepts that seem aimed at picking out a special kind of *connection* between things in the world. Causation is sometimes called, half jokingly, "the cement of the universe" (Mackie 1980; the phrase was first used by Hume [(1740) 1978]). In recent years, many philosophers have been skeptical about these concepts. But generally, their attitude has not been to reject them ("There is no such thing as causation") but instead to reconstrue these concepts in a very low-key way ("Yes, there is causation, but it is no more than this . . ."). In particular, philosophers have tried to analyze both laws and causation in terms of patterns in the *arrangement* of things, rather than some extra connection *between* things. Sometimes this project is referred to as "Humeanism," after David Hume, the first philosopher to develop a really focused suspicion about concepts of connection between events in nature (see also Lewis 1986b). The present-day Humeans do not have the same kind of empiricism as Hume, but they do want to avoid believing in any sort of unobservable cement connecting the universe together.

So a philosopher with Humean views will try to construe laws of nature as no more than regularities, or basic patterns, in the arrangement of events. To treat laws of nature in this way is to leave behind one of the familiar connotations of the term "law." Usually, we see laws as directing, or guiding, or governing in some way. It is possible (indeed traditional) to see laws of nature as governing the flow of events in the universe. Laws are seen as *responsible* for the regular patterns that we see, rather than being *identical* with those patterns (Dretske 1977; Armstrong 1983). The Humean regards this "governing" conception of laws as a seduction that must be avoided by hard-headed philosophers. The logical empiricist attitude toward laws of nature was basically Humean, in this sense.

The topic of causation has generated a similar debate. On one side we have those who see causation as basically some special kind of regular pattern in the flow of events. One the other side are those who see causation as a connection between things that is somehow *responsible* for the patterns (see Sosa and Tooley 1993). Perhaps this connection need not be seen as a mysterious philosophical entity; maybe it can be described by ordinary science (Dowe 1992; Menzies 1996).

For some years philosophers tended to discuss laws and their role in science in a way that had little contact with actual scientific work. In 1983 Nancy Cartwright delivered a wake-up call to the field with a book called *How the Laws of Physics Lie,* in which she argued that what people call "laws of physics" do not usually describe the behavior of real systems at all, but only describe the behavior of highly idealized fictional systems. Another important change that resulted from a closer look at actual science is that philosophers are no longer obsessed with natural laws as *the* goal of scientific theorizing. Over many years philosophers searched fields like biology for statements of laws of nature. Philosophers thought that any genuine science had to contain hypothesized laws and had to organize its ideas via the concept of a law. In fact, most biology has little use for the concept of a law of nature, but that does not make it any less scientific.

Further Reading

Wesley Salmon's *Four Decades of Scientific Explanation* (1989) is a very good survey of work on explanation between 1948 (the advent of the covering law theory) and the late 1980s. (The only thing marring Salmon's discussion is his rather eccentric theory about causation, which affects his treatment of explanation.)

A very good alternative discussion of causation and explanation can be found in Lewis 1986a. (Lewis's theory of causation is also eccentric. In fact, I guess *every* philosopher's theory of causation is eccentric; no two philosophers seem to agree. Lewis's discussion is compatible with a range of different views about causation, though.)

For unificationist theories of explanation, see Friedman 1974 and Kitcher 1989. These are fairly advanced papers.

Lewis discusses the Humean program in metaphysics in the preface to his 1986 *Philosophical Papers,* volume 2. Armstrong 1983 is a clear introduction to the more purely philosophical side of the literature on laws of nature. Beebee 2000 is a good discussion of the idea that laws "govern" things. Mitchell (2000) defends an interesting position on laws.

14

Bayesianism and Modern Theories of Evidence

14.1 New Hope

Through much of the twentieth century, the unsolved problem of confirmation hung over philosophy of science. What is it for an observation to provide evidence for, or confirm, a scientific theory? Back in chapter 3, I described how this problem was tackled by logical empiricism. The logical empiricists wanted to start from simple, obvious ideas—like the idea that seeing many black ravens confirms the hypothesis that all ravens are black—and build from there to an "inductive logic" that would help us understand testing in science. They failed, and we left the topic in a state of uncertainty and frustration.

Karl Popper was one person who could enjoy this situation, since he opposed the idea that confirmation is an essential part of science. After Popper, we launched into a discussion of historically oriented theories of science, like Kuhn's. These theories were not focused on the problem of confirmation in the way logical empiricism was. But the problem did not go away. Some philosophers continued to work on it, and even when it was not being discussed, it lurked in the background. If someone had come up with a really convincing theory of confirmation, it would have been harder to argue for radical views of the kind discussed in chapters 7–9. The absence of such a theory put empiricist philosophers on the defensive.

The situation has now changed. Once again a large number of philosophers have real hope in a theory of confirmation and evidence. The new view is called *Bayesianism.* The core ideas of this approach developed slowly through the twentieth century, but eventually these ideas started to look like they might actually solve the problem. The attitude of many is summed up in the title of a recent book by John Earman: *Bayes or Bust?* (1992). This title refers to a widespread feeling that this approach had *better* work, or philosophy of science might really be in trouble again.

Although Bayesianism is the most popular approach to solving these problems today, I am not in the Bayesian camp. Some parts of Bayesianism are undeniably powerful, but I would cautiously put my money on some different ideas. These will be introduced at the end of the chapter.

And before setting out on these topics, I should stress that this is the hardest chapter in the book. Some readers might want to jump to chapter 15.

14.2 Understanding Evidence with Probability

At this point I will shift my terminology. The term "confirmation" was used by the logical empiricists, but more recent discussions tend to focus on the concept of *evidence*. From now on I will follow this usage.

Bayesianism tries to understand evidence using probability theory. This idea is not new. It has often seemed natural to express some claims about evidence in terms of probability. Rudolf Carnap spent decades trying to solve the problem in this way. And outside philosophy this idea is familiar; we say that seeing someone's car outside a party makes it very *likely* that he is at the party. The mathematical fields of statistics and data analysis use probability theory to describe the kinds of conclusions that can be drawn from surveys and samples. And in law courts, we have become familiar with the description of forensic evidence, like DNA evidence, in terms of probability.

Consequently, many philosophers have tried to understand evidence using probability. Here is an idea that lies behind many of these attempts: when there is uncertainty about a hypothesis, observational evidence can sometimes *raise or lower the probability* of the hypothesis.

Bayesianism is one version of this idea. For Bayesians, there is a formula that is like a magic bullet for the evidence problem: *Bayes's theorem*. Thomas Bayes, an English clergyman, proved his theorem in the eighteenth century. As a theorem—as a piece of mathematics—his idea is very simple. But the Bayesians believe that Thomas Bayes struck gold.

Here is the magic formula in its simplest form:

$$(1) \quad P(h|e) = P(e|h)P(h)/P(e).$$

Here it is in a form that is more useful for showing how it works in philosophy of science:

$$(2) \quad P(h|e) = \frac{P(e|h)P(h)}{P(e|h)P(h) + P(e|not\text{-}h)P(not\text{-}h)}.$$

Here is how to read formulas of this kind: P(*X*) is the probability of *X*. P(*X*|*Y*) is the probability of *X* conditional upon *Y*, or the probability of *X* *given* *Y*.

How does this formula help us to understand confirmation of theories? Read "*h*" as a hypothesis and "*e*" as a piece of evidence. Then think of P(*h*) as the probability of *h* measured without regard for evidence *e*. P(*h*|*e*) is the probability of *h given e*, or the probability of the hypothesis *in the light of e*. Bayes's theorem tells us how to compute this latter number. As a consequence, we can measure what *difference* evidence *e* makes to the probability of *h*. So we can say that evidence *e confirms h* if P(*h*|*e*) > P(*h*). That is, *e* confirms *h* if it *makes h more probable* than it would otherwise be.

Picture someone changing her beliefs as evidence comes in. She starts out with P(*h*) as her assessment of the probability of *h*. If she observes *e*, what should her new view be about the probability of *h*? It seems that her new view of the probability of *h* should be given by P(*h*|*e*), which Bayes's theorem tells us how to compute. So Bayes's theorem tells us how to *update* probabilities in the light of evidence. (More on this updating later.)

Those are two central ideas in Bayesianism: the idea that *e confirms h* if *e* raises the probability of *h*, and the idea that probabilities should be *updated* in a way dictated by Bayes's theorem.

Bayes's theorem expresses P(*h*|*e*) as a function of two different kinds of probability. Probabilities of hypotheses of the form P(*h*) are called *prior* probabilities. Looking at formula 2, we see P(*h*) and P(*not-h*); these are the prior probability of *h* and the prior probability of the negation of *h*. These two numbers must add up to one. Probabilities of the form P(*e*|*h*) are often called "likelihoods," or the likelihoods of evidence on theory. In formula 2 we see two different likelihoods, P(*e*|*h*) and P(*e*|*not-h*). (These need not add up to any particular value.) Finally, P(*h*|*e*) is the "posterior probability" of *h*.

Suppose that all these probabilities make sense and can be known; let's see what Bayes's theorem can do. Imagine you are unsure about whether someone is at a party. The hypothesis that he is at the party is *h*. Then you see his car outside. This is evidence *e*. Suppose that before seeing the car, you think the probability of his going to the party is 0.5. And the probability of his car's being outside if he *is* at the party is 0.8, because he usually drives to such events, while the probability of his car's being outside if he is *not* at the party is only 0.1. Then we can work out the probability that he is at the party *given* that his car is outside. Plugging the numbers into Bayes's theorem, we get P(*h*|*e*) = (.5)(.8)/[(.5)(.8) + (.5)(.1)], which is almost 0.9. So seeing the car raises the probability of *h* from 0.5 to about 0.9; seeing the car strongly confirms the hypothesis that the person is at the party.

This all seems to be working well. We can do a lot with Bayes's theorem, *if* it makes sense to talk about probabilities in these ways. It is common to think that it is not too hard to interpret probabilities of the form P(*e*|*h*), the likelihoods. Scientific theories are supposed to tell us *what we are likely to see*. Some Bayesians underestimate the problems that can arise with this idea, but there is no need to pursue that yet. The probabilities that are more controversial are the prior probabilities of hypotheses, like P(*h*). What could this number possibly be measuring? And the posterior probability of *h* can only be computed if we have its prior probability. So although it would be good to use Bayes's theorem to discuss evidence, many interpretations of probability will not allow this because they cannot make sense of prior probabilities of theories. If we want to use Bayes's theorem, we need an interpretation of probability that will allow us to talk about prior probabilities. And that is what Bayesians have developed. This interpretation of probability is called the *subjectivist* interpretation.

14.3 The Subjectivist Interpretation of Probability

Most attempts to analyze probability have taken probabilities to measure some real and "objective" feature of events. A probability value is seen as measuring the *chance* of an event happening, where this chance is somehow a feature of the event itself and its location in the world. That is how we usually speak about the probabilities of horses winning races, for example. But according to the subjectivist interpretation, probabilities are *degrees of belief*. A probability measures a person's degree of confidence in the truth of some proposition. So if someone says that the probability the horse "Tom B" will win its race tomorrow is 0.4, he is saying something about *his degree of confidence* that the horse will win.

The subjectivist approach to probability was pioneered (independently) by two philosopher-mathematicians, Frank Ramsey and Bruno de Finetti, in the 1920s and 1930s. This interpretation of probability is not only important in philosophy; it is central to decision theory, which has great importance in the social sciences (especially economics). The majority of philosophers who want to use Bayes's theorem to understand evidence hold a subjectivist view of probability—at least for applications of probability theory to this set of problems, and sometimes more generally. Some are subjectivists because they feel they *have* to be in order to use Bayes's theorem; others think that subjectivism is the only interpretation of probability that makes sense anyway. The philosophical debates about Bayesianism also connect to debates about probability within mathematical statistics itself.

So let us look more closely at subjectivism about probability and how it relates to Bayesianism. The details of this topic are ferociously technical, but the main ideas are not too difficult.

Subjectivism sees probabilities as degrees of belief in propositions or hypotheses about the world. To find out what someone's degree of belief in a proposition is, we do not ask him or look inside his mind. Instead, we see his degrees of belief as revealed in his *gambling behavior,* both actual and possible. Your degrees of belief are revealed in which bets you would accept and which you would reject. Real people may be averse to gambling, even when they think the odds are good, or they may be prone to it even when the odds are bad. Here and in other places, Bayesianism seems to be treating not *actual* people but *idealized* people. But let us not worry too much about that. To read off a person's degrees of belief from his gambling behavior, we look for the odds on a given bet such that the person would be *equally willing to take either side* of the bet. Call these odds the person's *subjectively fair odds* for that bet. If we know a person's subjectively fair odds for a bet, we can read off his degree of belief in the proposition that the bet is about.

For example, suppose you think that 3:1 is fair for a bet on the truth of *h*. That is, a person who bets that *h* is true wins $1 if she is right and loses $3 if she is wrong. More generally, let us say that to bet *on h* at odds of X:1 is to be *willing to risk losing $X* if *h* is false, in return for a gain of $1 if *h* is true. So a large X corresponds to a lot of confidence in *h*. And if your subjectively fair odds for a bet on *h* are X:1, then your degree of belief in *h* is $X/(X + 1)$.

So far we have just considered one proposition, *h*. But your degree of belief for *h* will be related to your degrees of belief for other propositions as well. You will have a degree of belief for *h*&*j* as well, and for *not-h*, and so on. To find your subjective probability for *h*&*j*, we find your subjectively fair odds for a bet on *h*&*j*. So a person's belief system at a particular time can be described as a network of subjective probabilities. These subjective probabilities work in concert with the person's preferences ("utilities") to generate his or her behavior. From the Bayesian point of view, all of life is a series of gambles, in which our behavior manifests our bets about what the world is like.

Bayesians claim to give a theory of when a person's total network of degrees of belief is "coherent," or rational. They argue that a coherent set of degrees of belief has to follow the standard rules of the mathematics of probability.

Here is a quick sketch of these more technical ideas. Modern treatments of probability start from a set of axioms (most basic principles) first developed by the Russian mathematician Kolmogorov. Here is a version of those axioms that is used by subjectivists.

Axiom 1: All probabilities are numbers between 0 and 1 (inclusive).
Axiom 2: If a proposition is a tautology (trivially or analytically true), then it has a probability of 1.
Axiom 3: If h and h^* are exclusive alternatives (they cannot both be true), then $P(h\text{-or-}h^*) = P(h) + P(h^*)$.
Axiom 4: $P(h|j) = P(h\&j)/P(j)$, provided that $P(j) > 0$.

(Bayes's theorem is a consequence of axiom 4. $P(h\&j)$ can be broken down both as $P(h|j)P(j)$ and as $P(j|h)P(h)$. So these are equal to each other and Bayes's theorem follows trivially.)

Why should your degrees of belief follow these rules? Subjectivists argue for this with a famous form of argument called a "Dutch book." (My apologies to any readers who are Dutch.)

The argument is as follows: if your degrees of belief do *not* conform to the principles of the probability calculus, there are possible gambling situations in which you are *guaranteed* to lose money, no matter how things turn out. How can there be a guarantee? Because these situations are ones in which you are betting on *both* sides of a proposition (or betting on *all* the horses in the race), at various different odds. In these situations, if the degrees of belief you have do not conform to the probability calculus, and you are willing to accept any bet that fits with your degrees of belief, then you will be willing to accept combinations of bets that guarantee you a loss.

Here is a simple example involving a coin toss. Suppose your degree of belief that the toss will come out heads is 0.6 and your degree of belief that the toss will come out tails is 0.6. Then you have violated the probability calculus because, by axiom 3, P(heads or tails) = 1.2, which axiom 1 says is impossible. But suppose you persist with these degrees of belief and are willing to bet on them. Now suppose someone (a "Dutch bookie") offers you the following bets: (1) You are to bet $10 at 1.5:1 that the result will be heads, and (2) you are to bet $10 at 1.5:1 that the result will be tails.

You should accept both bets, because your subjectively fair odds for heads and for tails are both 1.5:1. (To go from a degree of belief p to odds of X:1, use $X = p/(1 - p)$.) But now you have accepted two bets that each pay worse than even money on the only two possible outcomes. So you are guaranteed to lose. If the coin lands heads, you win $10 on the heads bet but lose $15 on the tails bet, so you are $5 behind. The same applies—a net loss of $5—if the coin lands tails. You have fallen victim to a Dutch book. If you want to ensure that no one could possibly make a Dutch book against you, you must ensure that your degrees of belief follow the rules of probability theory. This is a simple case, but more complex arguments of the same kind can be used to show that *any* violation of the mathematical rules of probability makes a person vulnerable to a Dutch book.

Of course, there are not very many Dutch bookies out there, and one can avoid the threat by refusing to gamble at all. That is not the point. The point is supposed to be that the Dutch book argument shows that anyone who does not keep his degrees of belief in line with the probability calculus is irrational in an important sense.

Let us now connect these ideas to the problem of evidence. The ideas about belief and probability discussed in this section so far apply to a person's beliefs *at a specific time*. But we can use these ideas to give a theory of the rational *updating* of beliefs as evidence comes in. Bayes's theorem tells us about the relations between $P(h)$ and $P(h|e)$. Both those assignments of probability are made *before* e is observed. Then suppose e is actually observed. According to Bayesianism, the rational agent will update her degrees of belief so that her *new* overall confidence in h is derived from her *old* value of $P(h|e)$. So the key relationship in this updating process is

(3) $P_{new}(h) = P_{old}(h|e)$.

The probability $P_{new}(h)$ then becomes the agent's new *prior* probability for h, for use in assessing how to react to the *next* piece of evidence. So "today's posteriors are tomorrow's priors." A different set of Dutch book arguments is used to argue that a rational agent should do her updating of beliefs in accordance with formula 3. (Bayesianism has to make special moves to deal with "old evidence," evidence known *before* its relation to a hypothesis is assessed, and it also has a different formula to use when evidence e is itself uncertain.)

To finish this section (sigh of relief), note again that according to a subjectivist interpretation of probability, there is no way that one set of degrees of belief can be "closer to the real facts about probability" than another, so long as both sets of degrees of belief each follow the basic rules (axioms) of probability. At least, that is how things work according to *strict* subjectivism. Some Bayesians would like to recognize an objective sense of probability as well as a subjective sense, but this requires extra arguments.

14.4 Assessing Bayesianism

Bayesianism is an impressive set of ideas. There is a big literature on these topics, and I will not try to predict whether Bayesianism will work in the end. But many of the debates have to do with the role of prior probabilities, so they are worth further discussion.

In standard presentations of Bayesianism, a person is imagined to start out with an initial set of prior probabilities for various hypotheses, which are updated as evidence comes in. This initial set of prior probabilities is a sort of *free choice;* no initial set of prior probabilities is better than another so long as the axioms of probability are followed. This feature of Bayesianism is sometimes seen as a strength and sometimes as a weakness.

It can seem to be a weakness because Bayesianism cannot criticize very strange initial assignments of probability. And, one might think, where you end up after updating your probabilities must depend on where you start.

But this is only true in a sense. Bayesians argue that although prior probabilities are freely chosen and might be weird initially, the starting point gets "washed out" by incoming evidence, so long as updating is done rationally. The starting point matters less and less as more evidence is taken into account.

This idea is usually expressed as a kind of *convergence.* Consider two people with very different prior probabilities for *h* (neither probability exactly one or zero), and the same likelihoods for all possible pieces of evidence (e_1, e_2, e_3 . . .). And suppose the two people see all the same actual evidence. Then these two people's probability for *h* will get closer and closer. It can be proved that for *any* amount of initial disagreement about *h,* there will be *some* amount of evidence that will get the two people to any specified degree of closeness in their final probabilities for *h.* That is, if having final probabilities within (say) 0.001 of each other counts as being in close agreement, then no matter how far apart people start out, there is some amount of evidence that will get them within 0.001 of each other by the end. So initial disagreement is eventually washed out by the weight of evidence.

This convergence could, however, take a very long time. These "in the limit" proofs may not help much. As Henry Kyburg likes to put it, we must also accept that for *any* amount of evidence, and any measure of agreement, there is *some* initial set of priors such that this evidence will *not* get the two people to agree by the end. So some Bayesians have tried to work out a way of "constraining" the initial assignments of probability that Bayesianism allows.

I think there is a more basic problem with the arguments about convergence or the "washing out" of prior probabilities. The convergence proofs assume that when two people start with very different priors, they nonetheless agree about all their *likelihoods* (probabilities of the form $P(e_1|h)$, etc.). That is needed for disagreement about the priors to "wash out." But why should we expect this agreement about likelihoods? Why should two people who disagree massively on many things have the same likelihoods for all possible evidence? Why don't their disagreements affect their views on the *relevance* of possible observations? This agreement *might* be present, but there is no general reason why it should be. (This is another aspect of the problem of holism.) Presentations of Bayesianism often use simple examples involving gambling games or sampling processes, in which it seems that there will be agreement about likelihoods even when people have different priors. But those cases are not typical.

This argument suggests that convergence results do not help much with problems about theory choice in science. It's not clear whether this is a big problem for Bayesianism, however, as there is controversy about how important the "washing out of priors" really is to Bayesianism.

Prior probabilities are also the key to the standard Bayesian answer to Goodman's new riddle of induction. The riddle was introduced back in section 3.4. Suppose we are presented with two inductive arguments made from the same set of observations of green emeralds. One induction concludes that all emeralds are green, and the other concludes that all emeralds are grue. Why is one induction good and the other bad?

The standard Bayesian reply is that both inductive arguments are OK. Both hypotheses are confirmed by the observations of green emeralds. However, the "All emeralds are green" hypothesis will, for most people, have a much higher *prior* probability than the "All emeralds are grue" hypothesis. Then although both hypotheses are confirmed by the observations, the green-emerald hypothesis ends up with a much higher posterior probability than the grue-emerald hypothesis. That is the difference between the two inductions.

This does establish a difference, but *why* does the grue-emerald hypothesis have a low prior probability? Is anything stopping a person from having things the other way around—having a higher prior for the grue-emerald hypothesis? Nothing is preventing this. A person's prior probability for the grue-emerald hypothesis will usually be the result of much past experience with colors, minerals, and so on. But this *need* not be the case. Suppose you have never had a single thought about emeralds in your life, and you arbitrarily decide to set a higher prior for the grue-emerald hypothesis. Bayesianism offers no criticism of this decision, so long as your probabilities are internally coherent and you update properly. Is this a bad result, or a good one?

14.5 Scientific Realism and Theories of Evidence

Perhaps Bayesianism will win in the end. But in the rest of this chapter I will discuss some other ideas. I should note that it is not clear which of these ideas are really in competition with Bayesianism, as opposed to complementing it. My aim will be to tie the problem of evidence to the discussions of realism and naturalism in earlier chapters.

Let us look again at what a theory of evidence should try to analyze. Much twentieth-century empiricism based its discussions of evidence upon a simple picture of what scientific testing is supposed to achieve: the aim of testing is to confirm and disconfirm generalizations by means of observation. That is what is fundamental to science. In the case of *dis*confirmation, deductive logic will suffice. Confirmation was to be analyzed using a special inductive logic.

This view failed. It failed to connect with actual science, and it failed even in its own terms; it could not make much sense of confirmation *within* the simple picture being used.

So here is a better picture of what scientific testing aims to do. Testing in science is typically an attempt to choose between rival hypotheses about the hidden structure of the world. These hypotheses will sometimes be expressed using mathematical models, sometimes using linguistic descriptions, and sometimes in other ways. Sometimes the "hidden structures" postulated will be causal mechanisms, sometimes they will be mathematical relationships that may be hard to interpret in terms of cause and effect, and sometimes they will have some other form. Sometimes the aim is to work out a whole new kind of explanation, and sometimes it is just to work out the details (like the value of a key parameter). Sometimes the aim is to understand general patterns, and sometimes it is to reconstruct particular events in the past; I mean to include attempts to answer questions like "Where did HIV come from?" as well as questions like "Why do things fall when dropped?"

Back during the Scientific Revolution, it was common to think of the problem of evidence using the analogy of a clock. A scientist is like someone observing the motions of a clock from outside and trying to make inferences about the clock's inner workings (Shapin 1996). This analogy is too restricted as a picture of how science works, but it is closer to the truth than the picture found in much twentieth-century empiricist philosophy.

Approaching the problem of evidence with this realist orientation is a good idea, but we should be careful not to overgeneralize. I said that testing is *typically* an attempt to choose between hypotheses about hidden structure. *Typically* does not mean *always*. My discussion of realism in chapter 12 allowed that not all science is like this. Kuhn and Laudan have both emphasized, correctly, that different theories and paradigms can bring with them somewhat different accounts of what good scientific theories should do. These differences are likely to show up when we try to give a philosophical account of testing and evidence. For this reason and others, we may need to get used to the idea of a mixed or pluralist theory of evidence in science.

Once we move toward scientific realism, though, it becomes clear that understanding *explanatory inference* will be crucially important in any future account of testing. Explanatory inference, as defined in chapter 3, is inference from a set of data to a hypothesis about a structure or process that would explain the data. This is far more important in actual science than the philosopher's traditional conception of induction. Indeed, it can be argued that good inferences about what to expect, what patterns will continue in our experience, and so on, typically are developed *via* inferences about what the world is like and what processes are operating.

Philosophers have not made a lot of progress on understanding explanatory inference yet. But one idea that was long neglected, and recently revived, will surely turn out to be part of the answer. This is inference via the *elimination of alternatives,* supporting one option by ruling out others. I will call this "eliminative inference." (It is sometimes called "eliminative induction," but that is another overly broad use of the term "induction.")

Eliminative inference is, of course, the kind of reasoning associated with the famous fictional detective Sherlock Holmes. If we can rule out all the suspects except one, we know who committed the crime. This approach to evidence and testing has an odd history in twentieth-century philosophy. It was often neglected, partly because philosophers tended to assume that in science there is always an infinite number of possible alternatives to any given theory. If a theory has an infinite number of rivals, then ruling out any finite number of alternatives does not reduce the number of possibilities remaining. This argument might not be very good, however. Maybe there are ways of constraining the *relevant* alternatives to a theory being considered, in which case we might be able to rule out most or all of the relevant alternatives.

A chemist, John Platt, once wrote a paper in which he argued that good science is generally based on eliminative inference (1964). His view looked like a modified version of Popperian testing. The paper was mostly ignored by philosophers but taken seriously by quite a few scientists. In recent years, philosophers have begun to resurrect the idea of eliminative inference. For example, John Earman has done this within a Bayesian framework (1992), and Philip Kitcher has done so without linking the idea to Bayesianism (1993).

Eliminative inference can have a deductive or a nondeductive form. The simplest cases are those where we are able to decisively rule out all options except one. If we can do that, then our inference can be presented as a deductive argument. (As always, such an argument is only as good as its premises.) This is what Sherlock was trying to do. There are two ways in which a nondeductive element can be introduced. First, there may be a less decisive ruling out of alternatives; maybe we can only hope to show that all alternatives except one are very unlikely. Second, we need to consider the case where we are able to rule out most, but not all, of the alternatives to a hypothesis. Maybe as we rule out more and more of the alternatives to a given hypothesis, that hypothesis acquires a kind of partial support, even though some doubt remains. Perhaps we can say that the theory becomes more and more likely to be true. Clearly, in the nondeductive cases it might make sense to embed eliminative inference within a Bayesian framework, which enables us to handle the idea of probability in a precise way. This is indeed a compatible relationship, since Bayesianism explicitly handles evidence in a *comparative* manner (for one hypothesis to gain credibility, another hypothesis has to lose).

One important feature of eliminative inference is this: it is clear that scientists give arguments of this kind all the time; there is no possibility that this is just a philosophical fiction. Elimination is also clearly important in understanding the hardest cases for a theory of explanatory inference: those in which whole new *kinds* of explanations, models, or theories were established in science. For these often involve what were seen at the time as head-to-head competitions: Darwin versus nineteenth-century creationism, Galileo versus Aristotelian physics, Skinner's behaviorist theory of language versus the "cognitivist" approach of Chomsky. In fact, looking at the history of science reminds us of the chief difficulty with arguments of the eliminative form: how do we know that we have considered all the relevant alternatives? It can be argued that scientists constantly tend toward overconfidence on this point (Stanford 2001). Scientists often think they have ruled out (or rendered very unlikely) all the feasible alternatives to their preferred theory—but in hindsight we can see that in many cases they did not do so, because we now *believe* a theory they did not even *consider.* So a focus on eliminative inference has the potential to illuminate both the successes *and the failures* found in scientific reasoning.

An emphasis on eliminative inference will probably be part of any good future philosophical theory of testing and evidence in science. It should not be made too central, though. The idea that supporting one option *always* works by ruling out others is too narrow; there can also be more direct support of one option. An example is discussed in the next section.

I will mention one other crucial form of reasoning that we see in actual science. This one, however, is much more philosophically perplexing than eliminative inference. Scientists often support hypotheses via an appeal to *simplicity*, or "parsimony." This was discussed briefly in chapter 3. Given two possible explanations for the data, scientists often prefer the simpler one. Despite various elaborate attempts, I do not think we have made much progress on understanding the operation of, or justification for, this preference.

14.6 Procedural Naturalism (Optional Section)

In this section I will outline some of my own ideas on the topic of evidence and testing. These ideas are intended to provide an alternative general picture to Bayesianism. But some of the ideas can be (and have sometimes been) combined with Bayesianism. The general viewpoint described is also supposed to be compatible with the discussion of eliminative inference just above.

The main idea I will defend is that we should analyze evidence, confirmation, and testing by focusing on *procedures*. If an observation provides support for a theory, that will be by virtue of the procedure that the observation was embedded within. Not all procedures must be explicit, planned tests or experiments; some can be more informal.

This procedure-oriented view goes back a fair way. One important source is Hans Reichenbach, who did not follow standard logical positivist thinking about confirmation. Reichenbach was also influenced by some statistical methods used in science. My version of this idea will be linked to naturalism. It also uses the idea of *reliability;* a good procedure is one that has the capacity to reliably answer the questions we put to it. In order to have a simple label, I will call the view *procedural naturalism.*

I will illustrate this view by looking at a particular type of procedure employed often in science: using random samples to make inferences about the characteristics of a larger population. This is the kind of procedure involved when we use a survey to find out (for example) how many teenagers smoke cigarettes. In a way, this is the closest thing to a scientific home for the traditional philosophical picture of inductive inference. But it turns out that if we approach some of the standard philosophical problems from a procedure-based point of view, it makes a big difference. So let us look again at the two famous puzzle cases discussed in chapter 3, the ravens problem and Goodman's "grue" problem.

The ravens problem was described in section 3.3. If generalizations are confirmed by their instances, and if any observation that confirms *H* also confirms anything logically equivalent to *H*, then it seems that a white shoe confirms the hypothesis that all ravens are black. After all, the white shoe is a nonblack nonraven. So it is an instance of the generalization "all nonblack things are nonravens," which is logically equivalent to "all ravens are black."

In most philosophical discussions of this problem, there is little attention paid to how the observations of ravens (and shoes) are being collected. Of course, the whole example is very unrealistic. A biologist would not try to learn about bird color simply by generalizing from observed cases. But let us imagine that a biologist is doing something like this. Rather than relying on casual observation, though, the biologist uses a statistical method.

Let us now distinguish two questions about ravens.

The general raven question: What is the proportion of blackness among ravens?
The specific raven question: Is it the case that 100 percent of ravens are black?

Questions of this kind can be reliably answered using samples from a larger population, if we have a sample that is *random* and of a reasonable *size*. Statistical theory will tell us exactly how large a sample we need in order to get an answer with a desired degree of reliability (here "size" of sample means absolute size, not size in relation to the size of overall population). So how might we collect an appropriate sample?

The most obvious approach is to collect a random sample of *ravens* and record the birds' colors. A sample of this kind can be used to answer both the specific and the general raven questions, using ordinary statistical methods. So far, so good.

But now consider a more unusual approach. Suppose we could collect a random sample of *nonblack* things and record whether or not they are ravens. This method will be useless for answering the general raven question. Interestingly, though, it *can* be used to reliably answer the *specific* raven question. If there are nonblack ravens, we can learn this, in principle, by randomly sampling the nonblack things.

Now that we are imagining unusual sampling methods, there are two others to consider: collecting a sample of black things, and collecting a sample of nonravens. Neither of these can be used, without further assumptions, to answer either of the raven questions. Knowing what proportion of the black things are ravens does not tell us what proportion of ravens are black, and a sample of *non*ravens is of no use either.

So far we have distinguished between some procedures that can, and some procedures that cannot, answer our questions about ravens. Now we can look at the role of particular observations. Consider a particular observation of a white shoe. Does it tell us anything about raven color? *It depends on what procedure the observation was part of.* If the white shoe was encountered as part of a random sample of nonblack things, then it *is* evidence. It is just one data point, but it is a nonblack thing that turned out not to be a raven. It is part of a sample that we can use to answer the specific question (though not the general question), and work out whether there are nonblack ravens. But if the very same white shoe is encountered in a sample of nonravens, it tells us nothing. The observation is now part of a procedure that cannot answer either question.

The same is true with observations of black ravens. If we see a black raven in a random sample of ravens, it is informative. It is just one data point, but it is part of a sample that can answer our questions. But the same black raven tells us *nothing* about our two raven questions if it is encountered in a sample of black things; there is no way to use such a sample to answer either question. The role of procedures is fundamental; an observation is only evidence if it is embedded in the right kind of procedure. I think this is a very general fact about evidence and confirmation; Hempel was wrong to think that generalizations are always confirmed by observations of their instances. There is only confirmation (or support) if the underlying procedure was of the right kind. (Interestingly, this does not apply in the case of *deductive* relationships. A black raven refutes the hypothesis that *no* ravens are black, regardless of the procedure behind the observation. But deduction, as always, is special.)

That concludes my sketch of a solution to the ravens problem. It is a more elaborate version of the idea, discussed in chapter 3, that "order of observation" is important (Horwich 1982). But what is important is not *order,* but *procedures.*

I turn now to the grue problem (section 3.4). This is harder, because I believe that "the grue problem" actually combines several different problems together (including the very difficult problem of simplicity). But I will present part of an answer.

Let us continue thinking about inferences made from samples, using statistical methods. These methods can be very powerful, but they can only be used when some assumptions hold about the testing situation. One situation where these methods *cannot* be used, in their simple forms, is when the act of observing or collecting the data *changes* the particular objects being observed, in a way that is relevant to the question being asked. In some cases we might overcome the problem by taking into account the effects of our data collection and compensating for this fact. But special measures of some kind will be needed.

Now consider Goodman and his emeralds. Again, the philosophical literature has chosen a bad example here, but suppose we are making inferences about all emeralds by observing a random sample. This method would encounter a problem if the act of collecting or observing individual emeralds changed their color. In such a case, a simple extrapolation from the color of the sample to the color of the unobserved emeralds would be unreliable. This problem is obvious. But there is a less obvious connection between this case and the grue problem.

First, I should remind you that a grue object is *not* one that changes color at some special date. I am not trying to solve the grue problem by objecting to emeralds changing their color, or anything like that. A grue object is one that was first observed before 2010 and is green, *or* was not observed before 2010 and is blue. With that point clear, think about a sample of grue emeralds that we might have collected.

To keep things simple, suppose that *all* our previously observed emeralds are in the sample. So we have a big pile of emeralds, all of which are grue. The act of putting them in the sample did not physically change them, but something related is going on. If those particular emeralds had not been observed before 2010, they would not have been grue. After all, those emeralds are green, and anything green that was never observed before 2010 does not count as grue. So the grueness of an object *depends,* in an odd conceptual way, on whether or not the object has been observed before a certain date. Putting it loosely, the emeralds in the sample were *affected,* with respect to their grueness, by the fact that they have been observed before now. But that means we cannot *extrapolate* grueness from the sampled emeralds to the unsampled ones. We cannot do the extrapolation because the observation process has interfered—in an odd way—with the characteristics of the objects in the sample. This problem does not appear if we want to extrapolate *greenness* from a sample of emeralds; it only appears if we want to extrapolate grueness.

The grue problem (or this aspect of the grue problem) is a strange philosophical version of a familiar problem in statistical methodology in science. It is akin to what would be called a *confounding variable* problem. In a way, Goodman's term "grue" turns observation (or sampling) itself into a confounding variable. Frank Jackson (1975) proposed a solution to Goodman's problem of roughly this kind, but without tying the solution to statistical methods or the idea of a confounding variable. I follow up this idea in more detail in Godfrey-Smith (forthcoming). The idea I want to emphasize here is, again, the importance of focusing on *procedures* in thinking about evidence.

Further Reading

On Bayesianism, I find Colin Howson and Peter Urbach's *Scientific Reasoning: The Bayesian Approach* (1993) very helpful, though the book does not seem to be popular within the Bayesian camp. John Earman, *Bayes or Bust?* (1992), is for the technically minded. Skyrms, *Choice and Chance* (2000), is a classic introduction to probability and induction. Michael Resnik's *Choices* (1987) is a particularly helpful introduction to decision theory, subjective probability, and Dutch books.

Earman (1992, chap. 7) and Kitcher (1993, chap. 7) discuss and defend eliminative inference. (Neither of these is easy reading.) For one of the elaborate attempts to make sense of the scientific preference for simple theories, see Forster and Sober 1994.

The view that I call "procedural naturalism" amalgamates ideas from various sources. Reichenbach's main discussion is in *Experience and Prediction* (1938) and is presented more accessibly in *The Rise of Scientific Philosophy* (1951). Alvin Goldman's two big books *Epistemology and Cognition* (1986) and *Knowledge in a Social World* (1999) give a general treatment of epistemological questions emphasizing the *reliability* of methods, rules, and procedures. The technically minded might be interested in a recent area of work that could be seen as contributing to a procedural naturalist approach to explanatory inference. This is work on inference about *causal* structure in networks of interacting factors (Pearl 2000; Spirtes, Glymour, and Sheines 1993).

15

Empiricism, Naturalism, and Scientific Realism?

15.1 A Muddy Paste?

We now reach the end of the tour. The tour has covered nearly a century of argument about science, and it has visited some fairly extreme climates and landscapes along the way. I will finish the book by trying to tie together some of the various threads, hints, insights, and pieces of the puzzle that have emerged in the preceding chapters. In particular, I will connect three ideas: empiricism, naturalism, and scientific realism. These three "isms" have each been explored and, in some form, defended. The harder question is whether they can be combined into a package that makes sense as a whole. We can't just declare that we are all "empiricist naturalist realists" or "naturalistic realist empiricists" and consider the job done.

When I wrote the proposal for this book, publishers sent it out for comments. One anonymous reviewer reacted against the idea that at the end we would have a happy three-way marriage of empiricism, naturalism, and scientific realism. The reviewer saw these as three ideas that could each be defended fairly well individually but which do not go well together. There are conflicts between them, or at least between some of the pairs. To make a good case for scientific realism, for example, requires being *opposed* to some central ideas in the empiricist tradition. So the reviewer predicted that when the last chapter of the book tried to put the three ideas together, the result would be a "muddy paste."

This is a good image. We start with three sharp and distinctive colors, three different big ideas about science, but when we try to put them together, we get a mess. Or so the reviewer predicted. Despite this vivid warning, I will indeed try to put the three together in this chapter. Readers can decide for themselves whether the result is mud.

15.2 The Apparent Tensions

Empiricism traditionally holds that our source of knowledge about the world is experience. Naturalism holds that we can only hope to resolve philosophical problems (including epistemological problems) by approaching them within a scientific picture of ourselves and our place in the universe. Scientific realism holds that science can reasonably aim to describe the real structure of the world, including its unobservable structure. So why can't we believe all three of these at once? Where is the problem?

Much of the problem comes from the side of empiricism. I have several times summarized empiricism as the view that our only source of knowledge is experience. But this is, of course, a vague and indefinite idea, more a starting point than a philosophical position. When people have tried to fill out this idea, the result has often been a view with troublesome consequences.

Traditional empiricism was often beguiled by a picture of the mind as shut in behind a "veil of ideas," or sensations. If all we have access to is our sensory experience, what chance do we have of forming justified beliefs about what lies *beyond* the veil? The most extreme forms of empiricism have denied that it even makes sense to *talk* or *think* about what might lie beyond experience. And even from a less extreme empiricist point of view, it can be hard to see how *experience itself* could support a hypothesis about structures lying *behind* experience. Hence the temptation to see science as concerned only with patterns in experience itself, or patterns in the observable domain.

In recent years the tension between scientific realism and empiricism has often been debated under the heading "the underdetermination of theory by evidence." Empiricists argue that there will always be a range of alternative theories compatible with all our evidence. So we can never have good empirical grounds for choosing one of these theories over others and regarding it as representing how the world really is. If we have no empirical grounds for such a choice, then we have no grounds at all.

So much for empiricism and scientific realism. The other possible tensions are not so bad, but they are still worth discussing. First, the relation between empiricism and naturalism is not always harmonious, because empiricist philosophies have often had a *foundationalist* structure. For many empiricists, given that we only have direct access to our ideas and experiences, we must begin from that starting point when developing a philosophical theory of knowledge. But according to naturalism, the idea of "starting within the circle of our ideas and then working our way out" is a bad mistake.

Some have thought there is a different kind of tension between naturalism and empiricism. Sociologists of science (and others in neighboring fields) have held that the empiricist tradition in philosophy of science has been shown to be a collection of myths. If we look at how science actually works, we do not find experience acting as a neutral "arbiter" of theoretical disputes, in the way imagined by empiricism. Arguments about the theory-ladenness of observation (section 10.3) are often used to make this point. I tried to defuse most of these arguments, but they do continue to be influential.

The last possibility is tension between naturalism and scientific realism. Here we find less of a problem. Indeed, it is hard to be a naturalistic philosopher without taking science seriously as a description of the world; that suggests that naturalism *requires* a form of scientific realism. In general, there is indeed compatibility here, but there have also been some arguments given along similar lines to those in the previous paragraph. Sociologists of science have often viewed themselves as taking a properly naturalistic approach to science, in contrast with the philosophers' flights of fancy. So sociologists have sometimes argued that among the philosophical myths about science that we must abandon are myths about science's *contact with reality.*

So much for the possible tensions. The big one is between empiricism and realism. In the next section I will argue that the way to overcome this problem is *via* naturalistic ideas.

15.3 Empiricism Reformed

In this section I will describe a reformed version of empiricism. The argument will proceed in two steps. The first is a general philosophical discussion that has to do with epistemology in general. The second has to do with empiricism as a view about science.

As described above, empiricists (and many others) used to operate with a picture of the mind's access to the world that has been called the "veil of ideas" picture. The mind is seen as confined to its own sensations and thoughts, trying in vain to reach a hypothetical world beyond. Many philosophers now agree that this is a misleading picture. But it is easy to fall back into *relatives* of this view, both when thinking generally about the role of experience in guiding beliefs, and also in thinking about science. Philosophy of science has often hung onto *enough* of the old picture for trouble to arise. It is easy to fall back into a picture in which we distinguish two layers, or domains, in the world. One domain is accessible to us and familiar—the domain of experiences, or the domain of the observable. The other domain is inaccessible, mysterious, "theoretical," and problematic.

So how *should* we describe the role of experience? The right way to proceed is to cast empiricism within a naturalistic approach to philosophy. My version of this approach is influenced by the early-twentieth-century naturalism of John Dewey (1929).

From the naturalistic point of view, humans are biological organisms embedded in a physical world that we evolved to deal with. All our lives—including the most elaborate outgrowths of our social and intellectual lives—involve constant causal traffic and interaction with this world in which we are embedded. Our attempt to know about the world is just one aspect of our causal interaction with it; much of this interaction is more practical. Our perceptual mechanisms—eyes, ears, and so forth—are tools that we use to coordinate our dealings with the world. These mechanisms respond to physical stimuli caused by objects and events in our environments. From the inside, we can never establish with complete certainty what lies behind a particular sensory input. But looking at ourselves from "sideways-on," from the point of view taken by biology and psychology, we can establish regular principles concerning how our perceptual machinery responds to objects and events distant from us. We can work out how our perceptual machinery helps us to navigate the world.

So far this is not a point about science or even about human beings in particular. It is a claim about all the animals (and other organisms) that use perceptual mechanisms to adapt themselves to what is going on in their environments. But this point is enough to help us avoid some philosophical problems about our "access" to the world. We should not think in terms of two domains in reality, one accessible and one mysterious. We are biological systems embedded in a world containing objects of all sizes and at all different kinds of distance and remove from us. Our mechanisms of perception and action give us a variety of different kinds of contact with these objects. Our "access" to the world via thought and theory is really a complicated kind of causal interaction. This access to the world is constantly being expanded, as our technology improves. Parts of the world that must, at one time, be the subject of indirect and speculative inferences can later be much more directly observed, scanned, or assayed.

Think again about the problem of "underdetermination of theory by evidence." Empiricists have worried that there will always be rival alternative theories of the world that are equally compatible with our observations. Given this, how can we hope to have good knowledge of what the unobservable part of the world is really like? In thinking about this problem, return to the simpler case of perception itself. The same kind of issue arises here. Our perceptual mechanisms are used to form judgments about objects in the world around us, even though these mechanisms are only directly affected by stimuli like light and sound waves. In principle, there will always be alternative layouts of objects that could, in principle, give rise to the same stimuli affecting our senses. There is a kind of "underdetermination" here, as psychologists themselves often say. However, we *can* in fact make reliable judgments about what is around us, using perceptual mechanisms. And we can know that we are able to do this, by studying the operation of our perceptual mechanisms scientifically. In the case of perception, we can learn *what kind of reliability we have* in our attempts to know about the world.

The same sort of approach can be applied to inferences and modeling strategies in science itself. We can ask, What sort of reliability are we actually able to achieve, using different sorts of scientific reasoning and model-building strategies? Over time, structures and objects in the world can move from being so inaccessible that only speculative model-building can be applied to them, to being so accessible that their study is routine. Inferences about the genetic composition of an organism, for instance, have recently made this transition from being very indirect and cautious to being rather direct and routine, via DNA sequencing technologies. The constant shifting of these boundaries means that we can often go back and look at models developed when a structure was inaccessible. We can then ask, How well did we do? And more generally, Which approaches have tended to steer us toward good models, and which have tended to steer us toward bad ones? (Do scientific preferences for simple theories, for example, tend to lead to good choices?) To undertake this kind of investigation requires that we draw on work in the history of science. History tells us about actual cases where different approaches were tried and either succeeded or failed.

Let us now focus more closely on what makes science distinctive. Though humans all share their basic forms of contact with reality, as a consequence of their shared biological nature, there are huge differences in how different people and different intellectual cultures approach the problem of investigating and understanding the world. The fact that people all have their brains attached to sense organs at one end and behavioral mechanisms at the other end does not prevent them from disagreeing profoundly about the right way to learn about the world. One important point of disagreement is in how people handle the assessment of *big* ideas—big theories and explanatory hypotheses about the world—as opposed to how they handle everyday life. Maybe our shared biology is enough to make us all fairly empirical when we are trying to get food to eat and decide how to get home. But this does not apply to attempts to develop and justify explanatory theories about our overall place in the universe. Here we find sharp disagreements in approach, both within and across cultures.

At the end of chapter 10, I said that we might think of science as a something like a *strategy*. In this sense science is the strategy of subjecting even the biggest theoretical ideas, questions, and disputes to testing by means of observation. This strategy is not dictated to us by the nature of human language, the fundamental rules of thought, or our biology; it is more like a *choice*. The choice can be made by an individual or by a culture. The scientific strategy is to construe ideas, to embed them in surrounding frameworks, and to develop them, in such a way that exposure to experience is sought even in the case of the most general and ambitious hypotheses about the universe. That view of science is a kind of empiricism.

This description of science as a strategy is a start, but it needs to be made more precise. Back in chapter 1, I noted that there is a lot of variety in how the word "science" is used; there are very broad and very narrow uses of the term. Here I will outline a two-part story that is related to that "broad versus narrow" distinction. Let us distinguish the general scientific strategy from a particular way of organizing *how* the strategy is carried out. The strategy itself is the attempt to assess big ideas by exposing them to experience. In a broad sense, that is what science is all about. But the Scientific Revolution and the work that followed it also developed a particular, socially organized way of carrying out the strategy. The term "science" can also be used, more narrowly, to refer to that social organization.

I will say a bit more about how the two parts of the story fit together. I take from the empiricist tradition the idea of assessing ideas by exposing them to experience. An individual, all alone, can carry out a strategy of this kind. An individual could set up a private, self-contained program of formulating hypotheses and assessing them via observational testing. An individual can internalize the dialogue between the imaginative and critical voices. Such an individual might refuse to trust others and might attempt to get as close as possible to the old fantasy of the lone empiricist, relying entirely on his own experience.

This is a *possible* way of carrying out the scientific strategy, but obviously it is very far from the usual way. If our aim is to understand what makes the tradition of work deriving from the Scientific Revolution different from other approaches, then we need a different kind of story. We need to focus on the development and structure of a *socially organized way of carrying out* the basic scientific strategy.

The distinctive features of science as a social structure are found along two different dimensions. One has to do with the organization of work at a given time. Here we find the suggestion that science has developed a reward system and an internal culture that generate an efficient mixture of competition and cooperation, and a beneficial division of scientific labor across different approaches to a problem. These ideas were discussed in chapter 11. The general argument is that science (construed narrowly, as involving a particular social structure) is able to *coordinate the energies* of diverse individuals in an effective way.

The other dimension has to do with the relationships between different times, and with the transmission of ideas between scientific generations. The crucial feature we find along this dimension is that scientific work is *cumulative.* Each generation builds on the work of predecessors; current workers "stand on the shoulders" of earlier workers, as Isaac Newton once put it. This requires both trustworthy ways of transmitting ideas across time and (again) a reward system that makes it worthwhile to carry on where earlier workers left off.

With a social structure of this kind, the "dialogue between the imaginative and the critical voices" can become a real dialogue. We have social mechanisms in place that reliably bring about the checking and scrutinizing of ideas. To use a phrase suggested to me by Kim Sterelny, we get an "engine of self-correction" to accompany the speculative side of scientific thinking. In a situation like this, we can have a true division of labor in how the basic empiricist pattern is manifested. Some dogmatic and bloody-minded individuals can work within the system and even play a potentially useful role, provided that flexibility and open-mindedness is found in the community as a whole.

For several of the figures discussed in this book, the way that the empiricist strategy has been socially organized by modern science exhibits a remarkable *balance.* Or, more accurately, we seem to find a couple of different balances. One is a balance between *competition and cooperation;* this is, in a sense, the message of the work by Merton, Hull, and Kitcher discussed in chapters 8 and 11. The other is a balance between *criticism and trust.* That is one of the main themes of Kuhn's work. It is also part of the message of Shapin's work. Shapin would be reluctant to accept my description of the relation between criticism and trust as a "balance"; this term suggests that the relationship is a *good* one. Shapin does not take a stand on that issue. Kuhn, however, did think that the relationship between criticism and trust found in science is a uniquely effective one.

The idea of "balance," with its positive connotations, will make some people suspicious about this part of the story. The suspicion is understandable. When we describe the relations between competition and cooperation, and between criticism and trust, as exhibiting "balance," this makes those relationships sound like precious and fragile achievements. But why are we so sure that the present state is a good one? How do we know that we could not do *better* by changing the social organization of science? Feyerabend thought that recent science had *lost* its balance with respect to the relation between imaginative and pedestrian work, as we saw in chapter 7. In chapters 9 and 11, I also discussed the possibility of a feminist criticism of Hull's claim that the relation between competition and cooperation in science is well balanced.

Let us also look briefly at some historical issues. Once we have worked out which features of science's social organization are essential to our epistemological theory of science, we get some new historical questions to ask. Above I distinguished two dimensions of the social organization of science: the organization of work *at* a time and the organization of work *across* times. Were there crucial transitions that gave us these features, or did they evolve more gradually? Did they arise together or separately?

The cumulative structure of scientific work is something that is old in some fields and newer in others; this is something that can be gained and lost partially. In their historical description of the development of ideas about the universe from the ancient period to the early part of the modern, Toulmin and Goodfield emphasize the way that a cumulative structure was sometimes gained and then lost (1962). A sustained line of work would be set up by a "school," often in some particular city, and then it would fade and be replaced by a succession of individuals working alone, often "reinventing the wheel" over and over again. Gradually, though, and field by field, this haphazard pattern was replaced by more cumulative work.

Turning to the organization of scientific work *at* a time, and the relation between cooperation and competition in science, we find that the middle of the seventeenth century may be crucially important. Shapin and Schaffer (1985) emphasize the role of Robert Boyle and the Royal Society of London in setting up a new kind of culture of controlled criticism and a new kind of network of trust. This made possible new kinds of collaborative work. (Again, Shapin and Schaffer do not tell this story in a way that *endorses* this result, but many philosophers of science would want to do this.) If we regard this new organized culture of work as absolutely crucial to science, then the work of the "adventurers" of the earlier part of the Revolutionary period, like Galileo, becomes slightly less important in the story.

The case of alchemy is also interesting here. Alchemy was the precursor to chemistry, and it was influential through the end of the seventeenth century. (Newton was very interested in it.) Alchemy was a combination of practical work based on detailed recipes and an amazingly strange set of accompanying theories. (Rocks were seen as *growing*, in a quasi-biological sense, in the earth; chemical reactions were signified in astrological relationships between planets.) Alchemy was quite empirical in some ways—very results-oriented—but the work of alchemists was organized in a way that contrasts strikingly with modern science. Alchemy was often intensely secretive; rather than wide, accessible publication of results, there was a culture of private and restricted communication. This was partly because of the semi-mystical nature of the field, and partly because of the hope for massive financial benefits via finding a way to transmute other metals into gold. As Shapin and Schaffer emphasize, Robert Boyle contrasted his open, cooperative new scientific culture with the secrecy of the alchemists, as well as with the emptiness and dogmatism of Scholasticism.

In chapter 1, I raised the possibility that the fields and practices that we call "science" are too dissimilar for there to be a detailed philosophical "theory of science." (This possibility is one aspect of a recent discussion about the "disunity of science" [Galison and Stump 1996; Suppes 1981]). In several of the previous chapters, I have argued for "mixed" or "pluralist" views on particular issues. In chapter 7 I argued that some scientific fields might fit reasonably well with Kuhn's account of paradigm-dominated normal science (or something along the same lines), while others might fit better with the views of Laudan and Lakatos. My treatment of explanation in chapter 13 defended a "contextualist" position, and the same possibility arose also in my discussion of scientific realism. So the last half of this book has taken the idea of diversity in the nature of science fairly seriously, but this has not prevented the development of a philosophical account of those issues. Philosophy need not always strive for the most sweeping generalizations, and this kind of recognition of diversity need not involve relativism. Still, the account of the scientific strategy and its characteristic social structure described in this section is indeed rather general. Others might propose views that have less generality or unity than I am trying for here.

15.4 A Last Challenge

My discussion of the scientific strategy in the previous section was presented as a vindication of empiricism. But is this a misleading connection to make? (Some commentators on this book have thought so.) Has the story told in this book shown that empiricism was basically right all along, or has it *rejected* all the central ideas of the empiricist tradition?

This challenge can be posed by thinking again about Kuhn. We can see Kuhn as arguing that science cannot be described by any kind of simple empiricist formula, because science is a *much more complicated machine* than traditional empiricism ever imagined. Empiricist ideas are not just vague and incomplete; they get it wrong. Empiricist views have no resources to describe the complex *balances* found in scientific work, especially balances found in the social organization of science. Kuhn's view of science was the first place in this book where we encountered the possibility of a very complex theory of science of this kind. I criticized a number of the details of Kuhn's theory of science, but not in a way that seems likely to help us with that basic challenge. We seem likely to end up with just as complicated a view of science, or something even more so.

The empiricist can reply: "OK, there is a lot of complexity, but still the basic ideas of empiricism capture the most fundamental features of how science works. Let us not lose the wood for the trees!" That is indeed the reply that I propose. But we need to be aware of the objections to this position, as well as the points in its favor.

From the point of view of some of its opponents, empiricism is based on a hopelessly simple picture of what knowledge involves. Empiricism is often summarized by saying that the only source of knowledge is experience. But what is this talk of "sources" doing here? We ask, Is there just one source of knowledge, or more than one? This is like asking, Is there just one pipe leading into this tank, or more than one? But the process of learning about the world is not like that; epistemology is not plumbing.

So the critic of empiricism that I am imagining here is someone who thinks that the discussion of social structure, frameworks, rewards, and so on, in the middle part of this book should *replace* the simple empiricist ideas that we started out with. Suppose it can be shown that science works via a balance between competition and cooperation, and between criticism and trust. If that is the key to understanding science, it is *not* something that was suggested or summarized by traditional empiricist ideas. It is a different kind of story and also a better one. Or so says the critic.

There is a connection here to the problem of "objectivity" in science. Back in chapter 1, I noted that people often want to know whether science is objective; this is a central concept in many philosophical and sociological discussions of science. I said that I would avoid the term, because it is ambiguous and tends to set up the issues in a misleading way. Why is this? Let us look more closely at how the word is used. Sometimes people talking about objectivity have in mind a distinction, perhaps a vague one, between good and bad influences on belief. *Ob*jective influences on belief are contrasted with *sub*jective influences. And objectivity involves some kind of impartiality, or lack of bias. Perhaps it is accurate to say that "objectivity" is a term used to refer to a loose family of distinctions, each of which makes some sort of contrast between two ways of forming beliefs, one way that is dependent on caprice, prejudice, or point of view, and one that avoids such "subjective" influences.

At other times, the term "objectivity" is used to express a quite different idea. Some things *exist* objectively and some do not. Do colors exist objectively? Do moral values exist objectively? They are said to exist objectively if they exist independently of what people think of them. There is a link here to the issues about realism discussed in chapter 12.

In some discussions of science, these different senses of objectivity are brought together. Beliefs are said to be formed objectively when they are caused by, or guided by, real things. Science counts as objective if it is a process in which belief and theory change are controlled by contact with real things in the world. So *is* science objective in this sense? Or rather, when science is working properly, is it objective in this sense? Is the structure of science one that tends to produce objectivity?

The view of science defended in this book *sort of* says yes to these questions. But these are not good ways to ask the key questions; the concept of objectivity is unhelpful here. It is crude, and it tends to suggest false dichotomies. People find themselves asking, Are scientific belief and theory change controlled by real objects, or by social factors? Are scientific ideas the products of the real world, or are they products of human creativity? Are we responsible for what we know, or is the world responsible for it? (Shapin and Schaffer 1985, 344). These are all bad questions; they all involve false dichotomies. Scientific belief is not the product of us alone or of the world alone; it is the product of an interaction between our psychological capacities, our social organization, and the structure of the world. The world does not "stamp" beliefs upon us, in science or elsewhere. Still, science is *responsive* to the structure of the world, via the channel of observation.

The critic of empiricism that I am imagining here makes a similar kind of objection. Why hold onto old empiricist slogans, when they seem to set up the issues in such simplistic and misleading ways? Why not tell the story in entirely new terms? There is more in science than was dreamed of in the tired old empiricist tradition.

The critic of empiricism suspects that people like me want to hang onto empiricist ideas because they are pleasingly simple and often rhetorically useful. What makes science different from attempts to understand the world based on religious fundamentalism? When questions like this are asked, the empiricist seems able to give a simple and satisfying answer. "Science is different because it is a process in which beliefs are shaped by observation. Ideas are assessed not in terms of their origins, but in terms of how they stand up to testing. Science is open-minded, anti-authoritarian, and flexible." Nice and simple. Now suppose that these traditional empiricist ideas are replaced by a much more complex story, a story about delicate balances, a special reward system, moves within and between frameworks. . . . The defender of a more complex story might still insist that science is a superior approach to investigation. But the features that make science different will not be obvious, simple features, as they are according to the empiricist story. Simplicity is often attractive, but simple answers are often false.

So I do recognize the force of the argument that empiricism has been buried rather than reformed in the latter part of this book. Nonetheless, I think the argument is erroneous. It *does* "lose the wood for the trees," as the saying goes. Modern science involves both a general strategy and a complex social structure that carries out the strategy. The first part of this two-part account, as developed in this book, is a modified, naturalistic form of empiricism.

15.5 The Future

What are the key issues for philosophy of science in the near future? What should people work on? The problems grappled with in the preceding section are certainly worth further discussion. But I will end by mentioning three issues that follow up discussions in earlier chapters, which I think are especially interesting at the moment.

The first follows from the last section of chapter 7. What role do frameworks, paradigms, and similar constructs have in our understanding of theory change in science? Should we follow Kuhn and Carnap in having a sharp distinction between two "tiers" of conceptual change? Or is this a beguiling image that creates problems rather than solving them?

The second has to do with the reward system in science, and the relations between individual-level and community-level goals. So far the philosophical treatments of this topic have tended to generalize a lot, and it has been assumed that scientists have all internalized a similar set of motivations. Using input from sociology of science, it should be possible to tell a much more detailed story. What differences are there between different fields and different subcultures in science, for example? The relation between competition and cooperation in science is a fascinating topic.

The third follows from the last section of chapter 12. I used a broad concept of "representation" to describe the relation that science aims to achieve between theories and reality. I resisted the tendency to make concepts from the philosophy of language, like reference and truth, central in this part of the story. I also emphasized the role of models. But we do not have a good philosophical theory of representation yet, and even the most basic issues here are fraught with controversy. A big cloud of uncertainty still hangs over this part of philosophy of science.

Along with these clouds of uncertainty, however, I think we can see some fairly definite progress in the philosophy of science in recent years.

The idea that, in some way or other, all science is concerned with is the description of patterns in experience has finally been (mostly) abandoned. Scientific realism has been developed and defended in sophisticated forms. The field has become less dominated by questions about language, and proper attention is being paid to model-building as a crucial part of scientific work. Theories of testing and evidence are in vastly better shape than they were fifty years ago. The very idea of looking closely at the relation between the reward structure in science and epistemological issues is a crucial advance. So there has been progress, but there is still much to do.

Glossary

For definitions and quick discussions of other philosophical terms, I recommend Simon Blackburn's excellent book *The Oxford Dictionary of Philosophy.* Blackburn also gives more detail on many of the terms discussed here.

Sometimes confusion in newcomers to philosophy arises not from technicality, but from slightly different philosophical uses of everyday language. For example, in philosophy the word "strong" when applied to a view or a hypothesis does not mean *effective,* and it carries no positive (or negative) connotation. "Strong" means something more like *extreme, bold,* or *tendentious.* This is related to the use of the term in logic, where a strong claim is one that has a lot of implications. "Weak" in this sense means *cautious, hedged,* or *moderate.* Scientists sometimes uses "strong" in the same way. So a "strong" version of a view (empiricism, realism, etc.) is not necessarily better than a weak form. Confusingly, philosophers do sometimes say *strong argument* when they mean that the argument is good, or convincing.

After discussing each term below, I indicate the chapters or sections of the book in which the term is important. Terms in boldface have their own entry in the glossary.

There are two terms used in this book that are my own modifications of more standard terms. These are "explanatory inference" and "eliminative inference." In both cases I am avoiding overly broad use of the term "induction."

Abduction. One of the many terms for **explanatory inference**. This one was coined by C. S. Peirce. (3.2, 14.5)

Analytic/Synthetic Distinction. Analytic sentences are true or false simply in virtue of the meanings of the terms within them. Synthetic sentences are true or false in virtue of both the meanings of the words and the way the world is. **Logical positivism** treated this distinction as very important. Quine argued that it does not exist. (2.3, 2.4, 2.5)

Anomaly. In Kuhn's theory of science, an anomaly is a puzzle that resists solution by the methods of **normal science**. This is close to the word's ordinary meaning (roughly, something out of place). (5.4)

A Priori/A Posteriori Distinction. If something is known (or knowable) a priori, it is known (or knowable) independently of evidence gained via experience. Knowledge that relies on evidence from experience is a posteriori knowledge. (2.3)

Bayesianism. The theory of evidence and testing that gives a central role to *Bayes's theorem,* which is a provable result in probability theory. Bayesians treat all rational belief change as a matter of updating one's degrees of belief in propositions, in accordance with the principles of probability theory. (chapter 14)

Confirmation. A relationship of *support* between a body of evidence and a hypothesis or theory. Confirmation is not the same as proof; a theory can be highly confirmed and yet be false. **Logical positivism** and **logical empiricism** put much emphasis on the role of this relationship in science, usually trying to analyze it with an "inductive logic." Their attempts were not very successful. (chapters 3, 4, 14)

Constructivism (Social Constructivism, Metaphysical Constructivism). A word with many meanings. In the debates discussed in this book, "constructivism" usually refers to some sort of view in which knowledge (and sometimes, reality itself) is seen as *actively created* by human choices and social negotiation.

People advocating constructivist views often do not distinguish carefully between the view that *theories* (or classifications, or frameworks) are constructed, and the view that the *reality* described by those theories is constructed. I use the term "metaphysical constructivism" for views that explicitly claim that reality is in some sense constructed. (12.5).

Van Fraassen also uses the term "constructive empiricism" for his view of science, though his position has little in common with others standardly called constructivist. (12.6)

Corroboration. Popper used this term for something that a scientific theory acquires when it survives attempts to refute it. Sometimes this looks like another name for **confirmation**, which Popper rejected. (4.5) The term is also occasionally used (though not by Popperians) in a way that *is* roughly synonymous with confirmation or support.

Covering Law Theory. A theory of scientific explanation developed by the logical empiricists (see **logical empiricism**), especially Carl Hempel. The theory holds that to explain something is to show how to infer it in a good logical argument that includes a statement of a law of nature in the premises. (13.2)

Deductive Logic. The well-developed branch of logic dealing with patterns of argument that have the following feature: if the premises of the argument are true, then the conclusion is *guaranteed* to be true. This feature is called "deductive validity."

Deductive-Nomological Theory (D-N Theory). A term sometimes used for the **covering law theory** of explanation, although it only refers to some of the cases covered by that theory, the ones in which the argument used to explain something is a *deductive* argument.

Demarcation Problem. Popper's term for the problem of distinguishing scientific theories from nonscientific theories. (4.2, 4.6)

Eliminative Inference. A pattern of inference in which a hypothesis is supported by ruling out other alternatives. (Sometimes called "eliminative induction," even though these arguments can be deductively valid in some cases.) (14.5)

Empiricism. A diverse family of philosophical views, all asserting the fundamental importance of *experience* in explaining knowledge, justification, and rationality. A slogan used for traditional empiricism in this book is "Experience is the only source of real knowledge about the world." Not all empiricists would like that slogan. There are also empiricist theories of language, which connect the meanings of words with experience or some kind of observational testing. (chapters 2, 15, section 10.3)

Epistemology. The part of philosophy that deals with questions involving the nature of knowledge, the justification of beliefs, and rationality.

Explanandum. Whatever is *being explained,* in an explanation. (chapter 13).

Explanans. Whatever is *doing the explaining,* in an explanation. (chapter 13).

Explanatory Inference. An inference from a set of data to a hypothesis about a structure or process that would explain the data. There are many terms for this idea or ideas like it, including "abductive inference," "inference to the best explanation," "explanatory induction," and "theoretical induction."

In this book I treat this category as possibly overlapping with the category of **eliminative inference**. Some cases of explanatory inference might work via the elimination of alternative explanations. Others would treat these as two distinct categories. (3.2, 14.5)

Falsificationism. A view of science developed by Karl Popper. The word "falsificationism" can be used narrowly to refer to Popper's proposal for how to distinguish scientific theories from nonscientific theories (the **demarcation problem**). Falsificationism in this sense says that a theory is scientific if it has the potential to be refuted by some possible observation. The term is also used more broadly for Popper's view that all testing in science has the form of trying to refute theories by observation, and there is no such thing as the **confirmation** of a theory by passing observational tests. (chapter 4)

Foundationalism. A term used for theories that approach epistemological problems (see **epistemology**) by trying to show how human knowledge is built on a "foundation" of basic and completely certain beliefs. These might be beliefs about one's own current experiences, perhaps. (10.1, 10.2)

Holism. Holist arguments and positions can be found in many philosophical debates. Generally, a holist is someone who thinks that you cannot understand a particular thing without looking at its place in a larger whole. Two kinds of holism are important in this book. *Holism about testing* claims that we cannot test a single hypothesis or sentence in isolation. Instead, we can only test complex networks of claims and assumptions as wholes, because only these whole networks make definite predictions about what we should observe in a given situation. *Meaning holism* claims that the meaning of any word (or other expression) depends on its connections to every other expression in that language. (2.4, 2.5)

Hypothetico-Deductivism. This term can be used both for a method of doing science and for a more abstract view about **confirmation**. The hypothetico-deductive method (H-D method) is the most common description of good scientific procedure given in science textbooks. Versions of the method vary, but the basic steps are as follows. (1) Gather some observations, (2) formulate a hypothesis that would account for the observations, (3) deduce some new observational predictions from the hypothesis, and (4) see if those predictions are true. If they are true, go back to step 3. If they are false, regard the hypothesis as falsified and go back to step 2.

Some versions omit or alter step 1. Versions also differ on whether the scientist should regard the theory as confirmed if the predictions made by the hypothesis are true.

"Hypothetico-deductivism" is also used for a view about the nature of confirmation, as opposed to the procedures used in testing. Here the idea is that a hypothesis is confirmed when it can be used to derive true observational predictions.

Incommensurability. An important concept in Kuhn's and Feyerabend's theories of science. The basic idea is that different theories or paradigms can be *hard or impossible to compare,* in a properly unbiased way. For example, incommensurability about standards is the idea that different paradigms tend to bring with them slightly different standards for what counts as good evidence or good scientific work. If two paradigms bring different standards with them, which set of standards do we use if we want to choose between the two paradigms? Incommensurability about language holds that key scientific terms (like "mass," "force," etc.) can have different meanings in different paradigms. So in a sense, people within two different paradigms can be speaking slightly different languages, even if they seem to be using the same words. (6.3)

Induction. There are many senses of this term. One old sense refers to a method for doing science described by Francis Bacon in the seventeenth century. This method is usually described as one in which lots of particular facts should be gathered first, and generalizations and other hypotheses should be based on this stock of facts. (Bacon did not think that all science should follow this simple pattern.) In most of the discussions described in this book, this is *not* what "induction" means.

Instead, induction is a kind of argument, or pattern of inference, rather than a method or procedure.

I use "induction" for inferences in which particular cases are used to argue for a generalization that goes beyond the cases observed. So these arguments are not deductively valid (see **deductive logic**). The logical positivists and logical empiricists tended to use the term more broadly—for any inference that is not deductively valid but where the premises do support the conclusion to some extent. (chapters 3, 4, 14)

Instrumentalism. One kind of opposition to **scientific realism**. The main idea is that scientific theories should be seen as instruments used to predict observations, rather than as attempts to describe the real but hidden structures in the world that are responsible for the patterns found in observations. (12.4, 12.6)

Likelihoods. In **Bayesianism** and in statistics, "likelihood" is a technical term referring to the probability that something (*e*) will be observed, given the truth of some hypothesis (*h*). So likelihoods are probabilities of the form $P(e|h)$.

The term "likely" is often not tied to this technical meaning, though. Sometimes philosophers say "likely" just to mean probable and say "likelihood" just to mean probability. (14.2, 14.3, 14.4)

Logical Empiricism. I use this term for the more moderate views about knowledge, language, and science that derived from **logical positivism** and developed after World War II, especially in the United States. The term is sometimes used for logical positivism too, however (especially by those who think that not that much changed between the earlier and later stages). Logical empiricism was a scientifically oriented version of empiricism that emphasized the tools of formal logic. (2.5, chapter 3)

Logical Positivism. A novel, adventurous, and scientifically oriented form of **empiricism** that developed between the two world wars in Vienna, Austria. Sometimes known as **"logical empiricism,"** though I use this term for a later and more moderate development of the ideas. Leading figures were Moritz Schlick, Otto Neurath, and Rudolf Carnap. The view was based on developments in logic, philosophy of language, and philosophy of mathematics. The logical positivists famously dismissed a lot of traditional philosophy as meaningless. Early versions included the phenomenalist position (see **phenomenalism**) that all scientific claims could be translated into claims in a special language that referred only to observations. (chapter 2, 12.4)

Metaphysics. This term is usually now used to refer to a subfield within philosophy, which looks at a particular set of questions. These are general questions about the nature of reality itself, rather than (for example) how we know about this reality. Standard questions here include the nature of causation, the reality of the "external world," and the relation between mind and body.

The term is sometimes seen as referring to an investigation that goes beyond what can be addressed using science. Construed that way, metaphysics is regarded by many as a mistaken enterprise. (The logical positivists regarded most traditional metaphysical discussion as meaningless.) But in most current discussion, the term "metaphysics" refers to a set of questions and does not prejudge the right way to address them.

Model. A word with many senses, leading to frequent confusion. Sometimes "model" is used in science and philosophy of science just to mean a deliberately simplified theory. I generally follow another, narrower use of the term (especially in 12.7). In this sense, a model is a structure (either abstract or concrete) that is used to represent some other system. These are often, but by no means always, deliberately kept simple. The main "abstract" cases here are mathematical models used in science. In the "concrete" cases, one real physical system is used to represent another.

"Model" can also refer to an analogy that is used to *accompany* a theory and make it more comprehensible.

The term "model" also has a technical meaning in mathematical logic; here a model is a precise kind of interpretation of a set of sentences, one that treats the sentences as all true. This third sense has been used in philosophical attempts to formally analyze "the structure of theories," a project not discussed in this book (and about which I am skeptical).

Naturalism. An approach to philosophy that emphasizes the links (often, the "continuity") between philosophy and science. Naturalism is especially popular in epistemology and the philosophy of mind. Naturalism is sometimes taken to imply some sort of claim about the ultimately *physical* nature of everything that exists. So naturalists are thought to deny the existence of, for example, nonphysical souls. In this book I do *not* associate naturalism itself with any particular claims about what does and does not exist. For me, naturalism holds that the best way to address many philosophical problems is to approach them within our best current scientific picture of the world. (chapters 10, 11, 15, sections 12.3, 14.5)

Normal Science. In Kuhn's theory of science, normal science is the orderly form of science guided by a **paradigm**. Most science, for Kuhn, is normal science. A good normal scientist applies and does not usually question the fundamental ideas supplied by the paradigm. (5.3)

Objectivity. A term often used in a vague way to refer to beliefs or belief-forming procedures that avoid prejudice, caprice, and bias. The contrast is usually with "subjective" beliefs or procedures, which bear the influence of a particular point of view.

The term is also used to refer to a way in which things can be said to *exist;* something exists objectively if it exists independently of thought, language, or (again) a particular point of view.

The two meanings can be combined; objectivity in the sense of lack of bias might be seen as achieved through making beliefs responsive to the real world. (1.3, 15.4)

Operationalism (Operationism). A strongly empiricist view of science and scientific language developed by a physicist, Percy Bridgman, partly in response to Einstein's work in physics. According to operationalism, all good scientific language must either refer to observations or be definable in terms that refer only to observations. So this view is similar to **logical positivism** but is more a suggestion for how language *should* be used in science than a theory of meaning applied to all language. (2.3, 8.4)

Paradigm. A term made famous by Kuhn's theory of science. He used the term in a number of ways. I distinguish two main senses. In the *narrow* sense, a paradigm is an impressive achievement that inspires and guides a tradition of further scientific work—a tradition of **normal science.** In the *broad* sense, a paradigm is a whole "way of doing science" that has grown up around a paradigm in the narrow sense. In this sense, a paradigm will typically include theoretical ideas about the world, methods, and subtle habits of mind and standards used to assess "good work" in the field. (chapters 5, 6, section 7.7)

Pessimistic Induction from the History of Science (Pessimistic Meta-Induction). An argument against some forms of **scientific realism.** The argument holds that theories have changed so much in the history of science that we should not have much confidence in our current theories. In the past, scientists have often been very confident that their theories were true, but (the argument goes) they usually turned out to be wrong. So we should expect the same for our own current theories. The argument can also be made specifically about the reality of *entities* postulated by past and present theories. (12.3)

Phenomenalism. The view that when we seem to be talking and thinking about real physical objects, all we are really talking and thinking about are patterns in the flow of our sensations.

The word "phenomenon" is often used far more broadly than this strict meaning of "phenomenalism" would suggest. The word is used in much philosophy with something like its everyday meaning, that is (roughly), *something that happens.* In science, the term "phenomenological law" is sometimes used to refer to a law of nature that is, in some sense, not deeply explanatory but just describes a pattern or regularity.

Posterior Probability. In **Bayesianism,** a posterior probability is a probability of a hypothesis (*h*) given some piece of evidence (*e*). So it is a probability of the form P(*h*|*e*). (chapter 14)

Pragmatism. A family of unorthodox empiricist philosophical views that emphasize the relation between thought and action. For pragmatists, the chief purpose of thought and language is practical problem-solving. The "classical" figures in the movement are C. S. Peirce, William James, and John Dewey. Richard Rorty is a more recent defender of a form of pragmatism (but with less of a connection to empiricism). Pragmatists reject the correspondence theory of **truth.** (12.6)

Prior Probability. In **Bayesianism**, a prior probability is the initial or "unconditional" probability of a hypothesis (*h*), within an application of Bayes's theorem. So it is a probability of the form P(*h*). Bayes's theorem gives a formula for moving from the prior probability of a hypothesis to the **posterior probability**, the probability *given* some (usually new) piece of evidence. (chapter 14)

Rationalism. In an older usage of this term, rationalism holds that some real knowledge about the world can be gained via pure reasoning of a kind that does not depend on experience. Mathematics has been seen as an example. So in this sense, rationalism is opposed to **empiricism**.

More recently, the term has been used for vaguer ideas that do not necessarily clash with empiricism. As a view about science, "rationalism" is often used for the idea that theory change is guided by good reasoning and attention to evidence, as opposed to various kinds of bias or arbitrariness. For example, Popper's view of science, which I classify in this book as an unorthodox kind of empiricism, is often referred to as rationalist.

Realism. A huge variety of views can be described as "realist" in some sense or other, and debates about realism can involve many different issues and dimensions. Perhaps the most basic idea is this: a realist about X's is someone who thinks that X's exist in a way that does not depend on our thoughts, language, or point of view. Questions about realism can be asked very broadly, perhaps about *all* facts, or about ordinary objects in the physical world. They can also be asked more narrowly, in which case X's might be numbers, moral facts, colors, or some other special category.

This type of question is often recast as a question about language or about knowledge—what is the meaning of our term "X"? Is it a term that aims to designate some entity in the world? Can we ever have any knowledge at all about the alleged X's? A somewhat special set of issues arise in the case of ***scientific* realism**.

Relativism. The idea that the truth or justification of a claim, or the applicability of a standard or principle, depends on one's situation or point of view. Such a position can be asserted generally (about all truth or all standards) or specifically (about some particular domain, like morality or logic). The "point of view" might be that of an individual, a social group, the users of a particular language, or some other group. (6.3, 9.4, 9.5)

Research Program. In Lakatos's view of science, a research program is a sequence of scientific theories that all explore and develop the same basic theoretical ideas. Later theories in the sequence are developed in response to problems with the earlier ones. Some ideas in a research program—the "hard core"—are essential to the program and cannot be changed. Science typically involves ongoing competition between rival research programs in each field. (7.2)

Research Tradition. Laudan's research traditions are similar to Lakatos's **research programs**. There are some differences, however, and Laudan's concept is probably more useful. For example, Laudan's research traditions include more than just theoretical ideas about the world; they include values and methods as well. Also, for Laudan the borderline between the fundamental ideas of a research tradition and the ever-changing details is not necessarily fixed. (7.3)

Scientific Realism. A family of positions that assert some kind of realist (see **realism**) attitude toward the world as understood by science. I defend a fairly cautious kind of scientific realism. This version holds, roughly, that there is a real world that we all inhabit and that one reasonable goal of science is describing what the world is like.

Many other defenses of scientific realism include a general statement of confidence in our current scientific theories, or about progress in the history of science. Some also include detailed claims about scientific language. (2.5, chapter 12)

Subjectivism (also Personalism). An interpretation of the mathematics of probability theory, especially associated with **Bayesianism**. Subjectivists (at least of the strict kind) hold that probabilities are *degrees of belief* rather than measures of some kind of objective "chances" that exist in the world. Less strict versions of the view allow that there might be two kinds of probabilities, subjective ones and objective ones. (14.3)

Theory-Ladenness of Observation. A family of ideas that all claim, in some way, that observation cannot be an unbiased way to test rival theories (or larger units like paradigms), because observational judgments (or observation reports, or both) are affected by the theoretical beliefs of the observer. (10.3)

Truth. In ordinary discussion, a true claim or sentence is one that describes how things really are; a false claim is one that misrepresents the world. Some, but not all, philosophical treatments of truth follow this familiar idea.

"Correspondence" theories of truth hold that true statements are those that have some definite "matching" relationship to the world (so they obviously agree with the familiar, everyday view above). The term "correspondence" suggests a picturing of some kind, but this is not usually what is meant. It has been extremely difficult to say anything plausible about what this special relationship is. (I said "matching" just now, but that does not seem to help much either.) Other theories have tried to treat truth as depending only on what sort of *evidence* lies behind a claim or what sort of *usefulness* the claim has. More recently, some philosophers have argued that we should not think of truth as a special relationship to the world or a feature of a representation at all. Instead, we should think of the word "true" as a tool used in discussion to express agreement and to make some other harmless linguistic moves. (12.7)

Verificationism. A theory of meaning associated with **logical positivism**. Verificationism is often summarized with the claim that the meaning of a sentence is its method of verification. "Verification" here is a less appropriate word than "testing." Perhaps a better way to express the view is to say that to know the meaning of a sentence is the same thing as knowing how, in principle, to test it. The theory only applies to those parts of language that purport to describe the world (as opposed to expressing emotion, expressing commands, etc.). (2.3, 2.4)

References

Albert, David Z. 1992. *Quantum Mechanics and Experience.* Cambridge, MA: Harvard University Press.

Alvarez, Luis W., Walter Alvarez, Frank Asaro, and Helen V. Michel. 1980. Extraterrestrial Cause for the Cretaceous-Tertiary Extinction. *Science* 208:1095–1108.

Armstrong, David M. 1983. *What Is a Law of Nature?* Cambridge: Cambridge University Press.

———. 1989. *Universals: An Opinionated Introduction.* Boulder, CO: Westview Press.

Ayer, Alfred J. 1936. *Language, Truth, and Logic.* London: V. Gollancz.

Barkow, Jerome H., Leda Cosmides, and John Tooby, eds. 1992. *The Adapted Mind: Evolutionary Psychology and the Generation of Culture.* Oxford: Oxford University Press.

Barnes, Barry, and David Bloor. 1982. Relativism, Rationalism, and the Sociology of Knowledge. In *Rationality and Relativism,* edited by Martin Hollis and Steven Lukes. Cambridge, MA: MIT Press.

Barnes, Barry, David Bloor, and John Henry. 1996. *Scientific Knowledge: A Sociological Analysis.* Chicago: University of Chicago Press.

Beebee, Helen. 2000. The Non-Governing Conception of Laws of Nature. *Philosophy and Phenomenological Research* 61:571–94.

Biagioli, Mario, ed. 1999. *The Science Studies Reader.* New York: Routledge.

Bishop, Michael A. 1992. Theory-Ladenness of Perception Arguments. In *PSA 1992,* vol. 1., edited by David Hull, Micky Forbes, and Kathleen Okruhlik, 287–99. East Lansing, MI: Philosophy of Science Association.

Bishop, Michael A., and Stephen P. Stich. 1998. The Flight to Reference, or How Not to Make Progress in the Philosophy of Science. *Philosophy of Science* 65:33–49.

Bloor, David. 1976. *Knowledge and Social Imagery.* London: Routledge & Kegan Paul.

———. 1983. *Wittgenstein: A Social Theory of Knowledge.* London: Macmillan.

———. 1999. Anti-Latour. *Studies in the History and Philosophy of Science* 30:81–112.

Bridgman, Percy. 1927. The Operational Character of Scientific Concepts. In *The Logic of Modern Physics.* New York: Macmillan. Reprinted in *The Philosophy of*

Science, edited by Richard Boyd, Philip Gasper, and J. D. Trout (Cambridge, MA: MIT Press, 1991).

Bromberger, Sylvain. 1966. Why-Questions. In *Mind and Cosmos,* edited by R. Colodny. Pittsburgh: University of Pittsburgh Press.

Callebaut, Werner. 1993. *Taking the Naturalistic Turn, or, How Real Philosophy of Science Is Done.* Chicago: University of Chicago Press.

Campbell, Donald T. 1974. Evolutionary Epistemology. In *The Philosophy of Karl Popper,* edited by Paul Arthur Schilpp. La Salle, IL: Open Court.

Carnap, Rudolf. 1937. *The Logical Syntax of Language.* Translated by A. Smeaton. London: Routledge & Kegan Paul.

———. 1950. *Logical Foundations of Probability.* Chicago: University of Chicago Press.

———. 1956. Empiricism, Semantics, and Ontology. In *Meaning and Necessity: A Study in Semantics and Modal Logic,* by Rudolf Carnap. 2d ed. Chicago: University of Chicago Press.

———. 1995. *An Introduction to the Philosophy of Science.* Edited by M. Gardner. New York: Dover.

Carnap, Rudolf, Hans Hahn, and Otto Neurath. [1929] 1973. The Scientific Conception of the World: The Vienna Circle. In *Empiricism and Sociology,* edited by M. Neurath and R. S. Cohen. Dordrecht: Reidel.

Cartwright, Nancy. 1983. *How the Laws of Physics Lie.* Oxford: Oxford University Press.

Chalmers, Alan F. 1999. *What Is This Thing Called Science?* 3d ed. Indianapolis: Hackett.

Churchland, Paul M. 1988. Perceptual Plasticity and Theoretical Neutrality: A Reply to Jerry Fodor. *Philosophy of Science* 55:167–87.

Churchland, Paul M., and C. A. Hooker. 1985. *Images of Science: Essays on Realism and Empiricism.* Chicago: University of Chicago Press.

Cohen, I. Bernard. 1985. *The Birth of a New Physics.* New York: W. W. Norton.

Cohen, Robert Sonné, Paul K. Feyerabend, and Marx W. Wartofsky, eds. 1976. *Essays in Memory of Imre Lakatos.* Vol. 39 in *Boston Studies in the Philosophy of Science.* Dordrecht: Reidel.

Copernicus, Nicolaus. [1543] 1992. *On the Revolutions.* Translated by E. Rosen. Baltimore: Johns Hopkins University Press.

Darwin, Charles. [1859] 1964. *On the Origin of Species.* Edited by E. Mayr. Cambridge, MA: Harvard University Press.

Davidson, Donald. 1984. On the Very Idea of a Conceptual Scheme. In *Inquiries into Truth and Interpretation,* by Donald Davidson. Oxford: Clarendon Press.

Dear, Peter. 2001. *Revolutionizing the Sciences: European Knowledge and Its Ambitions, 1500–1700.* Princeton, NJ: Princeton University Press.

Dennett, Daniel C. 1978. *Brainstorms: Philosophical Essays on Mind and Psychology.* Montgomery, VT: Bradford Books.

———. 1995. *Darwin's Dangerous Idea: Evolution and the Meanings of Life.* New York: Simon & Schuster.

Devitt, Michael. 1997. *Realism and Truth.* 2d ed. Princeton, NJ: Princeton University Press.

Dewey, John. 1929. *Experience and Nature.* Rev. ed. La Salle, IL: Open Court.

———. 1938. *Logic: The Theory of Inquiry.* New York: H. Holt.

Doppelt, Gerald. 1978. Kuhn's Epistemological Relativism: An Interpretation and Defense. *Inquiry* 21:33–86.

Dowe, Philip. 1992. Process Causality and Asymmetry. *Erkenntnis* 37:179–96.

Downes, Stephen M. 1992. The Importance of Models in Theorizing: A Deflationary Semantic View. In *PSA 1992,* vol. 1., edited by David Hull, Micky Forbes, and Kathleen Okruhlik, 142–53. East Lansing, MI: Philosophy of Science Association.

———. 1993. Socializing Naturalized Philosophy of Science. *Philosophy of Science* 60:452–68.

Dretske, Fred I. 1977. Laws of Nature. *Philosophy of Science* 44:248–68.

———. 1988. *Explaining Behavior: Reasons in a World of Causes.* Cambridge, MA: MIT Press.

Dupré, John. 1993. *The Disorder of Things: Metaphysical Foundations of the Disunity of Science.* Cambridge, MA: Harvard University Press.

Earman, John. 1992. *Bayes or Bust? A Critical Examination of Bayesian Confirmation Theory.* Cambridge, MA: MIT Press.

Edmonds, David, and John Eidinow. 2001. *Wittgenstein's Poker: The Story of a Ten-Minute Argument between Two Great Philosophers.* New York: ECCO.

Eldredge, Niles, and Stephen J. Gould. 1972. Punctuated Equilibria: An Alternative to Phyletic Gradualism. In *Models in Paleobiology,* edited by T. J. Schopf. San Francisco: Freeman.

Feder, Kenneth L. 1996. *Frauds, Myths, and Mysteries: Science and Pseudoscience in Archaeology.* Mountain View, CA: Mayfield Publishers.

Feigl, Herbert. 1943. Logical Empiricism. In *Twentieth Century Philosophy,* edited by D. D. Runes. New York: Philosophical Library.

———. 1970. The "Orthodox" View of Theories: Remarks in Defense As Well As Critique. In *Theories and Methods of Physics and Psychology,* edited by Michael Radner and Stephen Winokur. Minnesota Studies in the Philosophy of Science, vol. 4. Minneapolis: University of Minnesota Press.

Feyerabend, Paul K. 1970. Consolations for the Specialist. In *Criticism and the Growth of Knowledge,* edited by Imre Lakatos and Alan Musgrave. Cambridge: Cambridge University Press.

———. 1975. *Against Method: Outline of an Anarchistic Theory of Knowledge.* Atlantic Highlands, NJ: Humanities Press.

———. 1978. *Science in a Free Society.* London: New Left Books.

———. 1981. *Philosophical Papers.* Cambridge: Cambridge University Press.

Fine, Arthur. 1984. The Natural Ontological Attitude. In *Scientific Realism,* edited by Jarrett Leplin. Berkeley: University of California Press.

Fodor, Jerry A. 1981. *Representations: Philosophical Essays on the Foundations of Cognitive Science.* Cambridge, MA: MIT Press.

———. 1983. *The Modularity of Mind.* Cambridge, MA: MIT Press.

———. 1984. Observation Reconsidered. *Philosophy of Science* 51:23–43.

Fodor, Jerry A., and Ernest LePore. 1992. *Holism: A Shopper's Guide*. Oxford: Blackwell.

Forster, Malcolm R., and Elliott Sober. 1994. How to Tell When Simpler, More Unified, or Less Ad Hoc Theories Will Provide More Accurate Predictions. *British Journal for the Philosophy of Science* 45:1–35.

Friedman, Michael. 1974. Explanation and Scientific Understanding. *Journal of Philosophy* 71:5–19.

———. 1999. *Reconsidering Logical Positivism*. New York: Cambridge University Press.

———. 2000. *A Parting of the Ways: Carnap, Cassirer, and Heidegger*. Chicago: Open Court.

———. 2001. *Dynamics of Reason, Stanford Kant Lectures*. Stanford, CA: CSLI Publications.

Futuyma, Douglas J. 1998. *Evolutionary Biology*. 3d ed. Sunderland, MA: Sinauer Associates.

Galileo Galilei. [1623] 1990. The Assayer. In *Discoveries and Opinions of Galileo,* translated by Stillman Drake. New York: Anchor Books.

———. [1632] 1967. *Dialogue concerning the Two Chief World Systems, Ptolemaic & Copernican*. Berkeley: University of California Press.

Galison, Peter. 1990. Aufbau/Bauhaus: Logical Positivism and Architectural Modernism. *Critical Inquiry* 16:709–52.

———. 1997. *Image and Logic: A Material Culture of Microphysics*. Chicago: University of Chicago Press.

Galison, Peter, and David J. Stump, eds. 1996. *The Disunity of Science*. Stanford, CA: Stanford University Press.

Garrett, Don, and Edward Barbanell. 1997. *Encyclopedia of Empiricism*. Westport, CT: Greenwood Press.

Giere, Ronald N. 1988. *Explaining Science: A Cognitive Approach*. Chicago: University of Chicago Press.

Giere, Ronald N., and Alan W. Richardson, eds. 1997. *Origins of Logical Empiricism*. Vol. 16 in Minnesota Studies in the Philosophy of Science. Minneapolis: University of Minnesota Press.

Glymour, Clark N. 1980. *Theory and Evidence*. Princeton, NJ: Princeton University Press.

Godfrey-Smith, Peter. 1996. *Complexity and the Function of Mind in Nature*. Cambridge: Cambridge University Press.

———. Forthcoming. Goodman's Problem and Scientific Methodology. *Journal of Philosophy*.

Goldman, Alvin I. 1986. *Epistemology and Cognition*. Cambridge, MA: Harvard University Press.

———. 1999. *Knowledge in a Social World*. Oxford: Oxford University Press.

Good, I. J. 1967. The White Shoe Is a Red Herring. *British Journal for the Philosophy of Science* 17:322.

Goodman, Nelson. 1955. *Fact, Fiction & Forecast.* Cambridge, MA: Harvard University Press.

———. 1972. *Problems and Projects.* Indianapolis: Bobbs-Merrill.

———. 1978. *Ways of Worldmaking.* Indianapolis: Hackett.

———. 1996. Starmaking. In *Starmaking: Realism, Anti-Realism, and Irrealism,* edited by P. McCormick. Cambridge, MA: MIT Press.

Gould, Stephen J. 1977. Eternal Metaphors of Paleontology. In *Patterns of Evolution as Illustrated by the Fossil Record,* edited by A. Hallam. New York: Elsevier Scientific.

———. 1980. Is a New and General Theory of Evolution Emerging? *Paleobiology* 6:119–30.

———. 2002. *The Structure of Evolutionary Theory.* Cambridge, MA: Harvard University Press.

Gregory, Richard L. 1970. *The Intelligent Eye.* New York: McGraw-Hill.

Gross, Paul R., and N. Levitt. 1994. *Higher Superstition: The Academic Left and Its Quarrels with Science.* Baltimore: Johns Hopkins University Press.

Hacking, Ian. 1983. *Representing and Intervening: Introductory Topics in the Philosophy of Natural Science.* Cambridge: Cambridge University Press.

Haraway, Donna. 1989. *Primate Visions: Gender, Race, and Nature in the World of Modern Science.* New York: Routledge.

Harding, Sandra G. 1986. *The Science Question in Feminism.* Ithaca, NY: Cornell University Press.

———. 1996. Rethinking Standpoint Epistemology: What Is "Strong Objectivity"? In *Feminism and Science,* edited by E. F. Keller and H. E. Longino. Oxford: Oxford University Press.

Harman, Gilbert H. 1965. Inference to the Best Explanation. *Philosophical Review* 74:88–95.

Harvey, David. 1989. *The Condition of Postmodernity: An Enquiry into the Origins of Cultural Change.* Oxford: Blackwell.

Heilbroner, John. 1999. *The Worldly Philosophers: The Lives, Times, and Ideas of the Great Economic Thinkers.* 7th ed. New York: Touchstone Books.

Hempel, Carl G. 1958. The Theoretician's Dilemma. In *Concepts, Theories, and the Mind-Body Problem,* edited by Herbert Feigl, Michael Scriven, and Grover Maxwell. Minnesota Studies in the Philosophy of Science, vol. 2. Minneapolis: University of Minnesota Press.

———. 1965. *Aspects of Scientific Explanation and Other Essays in the Philosophy of Science.* New York: Free Press.

———. 1966. *Philosophy of Natural Science.* Englewood Cliffs, NJ: Prentice-Hall.

Hempel, Carl G., and Paul Oppenheim. 1948. Studies in the Logic of Explanation. *Philosophy of Science* 15:135–75.

Henry, John. 1997. *The Scientific Revolution and the Origins of Modern Science.* New York: St. Martin's Press.

Hesse, Mary B. 1966. *Models and Analogies in Science.* Notre Dame, IN: University of Notre Dame Press.

Hobbes, Thomas. [1660] 1996. *Leviathan.* Edited by J. C. A. Gaskin. Oxford: Oxford University Press.

Hollis, Martin, and Steven Lukes, eds. 1982. *Rationality and Relativism.* Cambridge, MA: MIT Press.

Horgan, John. 1996. *The End of Science: Facing the Limits of Knowledge in the Twilight of the Scientific Age.* Reading, MA: Addison-Wesley.

Horwich, Paul. 1982. *Probability and Evidence.* Cambridge: Cambridge University Press.

———. 1990. *Truth.* Oxford: Blackwell.

———, ed. 1993. *World Changes: Thomas Kuhn and the Nature of Science.* Cambridge, MA: MIT Press.

Howson, Colin, and Peter Urbach. 1993. *Scientific Reasoning: The Bayesian Approach.* 2d ed. Chicago: Open Court.

Hoyningen-Huene, Paul. 1993. *Reconstructing Scientific Revolutions: Thomas S. Kuhn's Philosophy of Science.* Chicago: University of Chicago Press.

Hrdy, Sarah Blaffer. 1999. *The Woman That Never Evolved.* Rev. ed. Cambridge, MA: Harvard University Press.

———. 2002. Empathy, Polyandry, and the Myth of the Coy Female. In *The Gender of Science,* edited by Janet A. Kourany. Upper Saddle River, NJ: Prentice Hall.

Hull, David L. 1988. *Science as a Process: An Evolutionary Account of the Social and Conceptual Development of Science.* Chicago: University of Chicago Press.

———. 1999. The Use and Abuse of Sir Karl Popper. *Biology and Philosophy* 14:481–504.

Hume, David. [1739] 1978. *A Treatise of Human Nature.* Edited by L. A. Selby-Bigge and P. H. Nidditch. Oxford: Oxford University Press.

———. [1740] 1978. An Abstract of *A Treatise of Human Nature.* In *A Treatise of Human Nature,* edited by L. A. Selby-Bigge and P. H. Nidditch. Oxford: Oxford University Press.

Huntington, Samuel P. 1996. *The Clash of Civilizations and the Remaking of World Order.* New York: Simon & Schuster.

Jackson, Frank. 1975. Grue. *Journal of Philosophy* 72:113–31.

Kant, Immanuel. [1781] 1998. *Critique of Pure Reason.* Translated by P. Guyer and A. W. Wood. Cambridge: Cambridge University Press.

Kauffman, Stuart A. 1993. *The Origins of Order: Self-Organization and Selection in Evolution.* Oxford: Oxford University Press.

Keller, Evelyn Fox. 1983. *A Feeling for the Organism: The Life and Work of Barbara McClintock.* San Francisco: W. H. Freeman.

———. 2002. A World of Difference. In *The Gender of Science,* edited by Janet A. Kourany. Upper Saddle River, NJ: Prentice Hall.

Keller, Evelyn Fox, and Helen E. Longino, eds. 1996. *Feminism and Science.* Oxford: Oxford University Press.

Kimura, Motoo. 1983. *The Neutral Theory of Molecular Evolution.* Cambridge: Cambridge University Press.

Kitcher, Philip. 1978. Theories, Theorists, and Theoretical Change. *Philosophical Review* 87:519–47.

———. 1981. Explanatory Unification. *Journal of Philosophy* 48:507–31.

———. 1989. Explanatory Unification and the Causal Structure of the World. In *Scientific Explanation,* edited by Philip Kitcher and Wesley Salmon. Minnesota Studies in the Philosophy of Science, vol. 16. Minneapolis: University of Minnesota Press.

———. 1990. The Division of Cognitive Labor. *Journal of Philosophy* 87:5–22.

———. 1992. The Naturalists Return. *Philosophical Review* 101:53–114.

———. 1993. *The Advancement of Science.* Oxford: Oxford University Press.

———. 2001. *Science, Truth, and Democracy.* Oxford: Oxford University Press.

Koertge, Noretta, ed. 1998. *A House Built on Sand: Exposing Postmodernist Myths about Science.* New York: Oxford University Press.

Koestler, Arthur. 1968. *The Sleepwalkers.* New York: Macmillan.

Kornblith, Hilary. 1993. *Inductive Inference and Its Natural Ground: An Essay in Naturalistic Epistemology.* Cambridge, MA: MIT Press.

———, ed. 1994. *Naturalizing Epistemology.* 2d ed. Cambridge: MIT Press.

Kourany, Janet A., ed. 2002. *The Gender of Science.* Upper Saddle River, NJ: Prentice Hall.

Kuhn, Thomas S. 1957. *The Copernican Revolution: Planetary Astronomy in the Development of Western Thought.* Cambridge, MA: Harvard University Press.

———. 1970. Reflections on My Critics. In *Criticism and the Growth of Knowledge,* edited by Imre Lakatos and Alan Musgrave. Cambridge: Cambridge University Press.

———. 1977a. Concepts of Cause in the Development of Physics. In *The Essential Tension: Selected Studies in Scientific Tradition and Change,* by Thomas S. Kuhn. Chicago: University of Chicago Press.

———. 1977b. *The Essential Tension: Selected Studies in Scientific Tradition and Change.* Chicago: University of Chicago Press.

———. 1977c. Objectivity, Value Judgment, and Theory Choice. In *The Essential Tension: Selected Studies in Scientific Tradition and Change,* by Thomas S. Kuhn. Chicago: University of Chicago Press.

———. 1978. *Black-Body Theory and the Quantum Discontinuity, 1894–1912.* Oxford: Oxford University Press.

———. 1996. *The Structure of Scientific Revolutions.* 3d ed. Chicago: University of Chicago Press. The first edition was published in 1962.

———. 2000. *The Road since Structure: Philosophical Essays, 1970–1993, with an Autobiographical Interview.* Edited by James Conant and John Haugeland. Chicago: University of Chicago Press.

Lakatos, Imre. 1970. Falsification and the Methodology of Scientific Research Programmes. In *Criticism and the Growth of Knowledge,* edited by Imre Lakatos and Alan Musgrave. Cambridge: Cambridge University Press.

———. 1971. History of Science and Its Rational Reconstructions. In *PSA 1970,* edited by Roger C. Buck and Robert S. Cohen, 91–136. Dordrecht: Reidel.

Lakatos, Imre, and Alan Musgrave, eds. 1970. *Criticism and the Growth of Knowledge.* Cambridge: Cambridge University Press.

Langton, Christopher G., Charles Taylor, J. Doyne Farmer, and Steen Rasmussen, eds. 1992. *Artificial Life II.* Proceedings of the Workshop on Artificial Life, February 1990, Santa Fe, NM. Redwood City, CA: Addison-Wesley.

Latour, Bruno. 1987. *Science in Action: How to Follow Scientists and Engineers through Society.* Cambridge, MA: Harvard University Press.

———. 1988. *The Pasteurization of France.* Cambridge, MA: Harvard University Press.

———. 1993. *We Have Never Been Modern.* Cambridge, MA: Harvard University Press.

Latour, Bruno, and Steve Woolgar. 1979. *Laboratory Life: The Construction of Scientific Facts.* Beverly Hills, CA: Sage Publications. Reprint, Princeton, N.J.: Princeton University Press, 1986.

Laudan, Larry. 1977. *Progress and Its Problems: Toward a Theory of Scientific Growth.* Berkeley: University of California Press.

———. 1981. A Confutation of Convergent Realism. *Philosophy of Science* 48:19–48.

———. 1987. Progress or Rationality? The Prospects for Normative Naturalism. *American Philosophical Quarterly* 24:19–31.

Leplin, Jarrett, ed. 1984. *Scientific Realism.* Berkeley: University of California Press.

Levy, Steven. 1992. *Artificial Life: The Quest for a New Creation.* New York: Pantheon Books.

Lewis, David. 1983. New Work for a Theory of Universals. *Australasian Journal of Philosophy* 61:343–77.

———. 1986a. Causation and Explanation. In *Philosophical Papers,* by David Lewis, vol. 2. Oxford: Oxford University Press.

———. 1986b. Introduction to *Philosophical Papers,* by David Lewis, vol. 2. Oxford: Oxford University Press.

Lingua Franca Editors. 2000. *The Sokal Hoax: The Sham That Shook the Academy.* Lincoln: University of Nebraska Press.

Lipton, Peter. 1991. *Inference to the Best Explanation.* London: Routledge.

Lloyd, Elisabeth A. 1993. Pre-Theoretical Assumptions in Evolutionary Explanations of Female Sexuality. *Philosophical Studies* 69:139–53.

———. 1997. Feyerabend, Mill, and Pluralism. *Philosophy of Science* 64 (4): S396–S408.

Lloyd, Genevieve. 1984. *The Man of Reason: "Male" and "Female" in Western Philosophy.* Minneapolis: University of Minnesota Press.

Longino, Helen E. 1990. *Science as Social Knowledge: Values and Objectivity in Scientific Inquiry.* Princeton, NJ: Princeton University Press.

Lynch, Michael, ed. 2001. *The Nature of Truth.* Cambridge, MA: MIT Press.

Lyotard, Jean-Francois. 1984. *The Postmodern Condition: A Report on Knowledge.* Translated by Geoff Bennington and Brian Massumi. Minneapolis: University of Minnesota Press.

Mach, Ernst. 1897. *Contributions to the Analysis of the Sensations.* Translated by C. M. Williams. Chicago: Open Court.

MacKenzie, Donald A. 1981. *Statistics in Britain, 1865–1930: The Social Construction of Scientific Knowledge.* Edinburgh: Edinburgh University Press.

Mackie, J. L. 1980. *The Cement of the Universe: A Study of Causation.* 2d ed. Oxford: Oxford University Press.

Masterman, Margaret. 1970. The Nature of a Paradigm. In *Criticism and the Growth of Knowledge,* edited by Imre Lakatos and Alan Musgrave. Cambridge: Cambridge University Press.

Maxwell, Grover. 1962. The Ontological Status of Theoretical Entities. In *Scientific Explanation, Space, and Time,* edited by Herbert Feigl and Grover Maxwell. Minnesota Studies in the Philosophy of Science, vol. 3. Minneapolis: University of Minnesota Press.

McMullin, Ernan. 1984. A Case for Scientific Realism. In *Scientific Realism,* edited by Jarrett Leplin. Berkeley: University of California Press.

Menzies, Peter. 1996. Probabilistic Causality and the Pre-Emption Problem. *Mind* 105:85–117.

Merton, Robert K. [1957] 1973. Priorities in Scientific Discovery. In *The Sociology of Science: Theoretical and Empirical Investigations,* edited by Norman Storer. Chicago: University of Chicago Press.

———. 1973. *The Sociology of Science: Theoretical and Empirical Investigations.* Edited by Norman Storer. Chicago: University of Chicago Press.

Mill, John Stuart. [1859] 1978. *On Liberty.* Edited by E. Rapaport. Indianapolis: Hackett.

———. 1865. *An Examination of Sir William Hamilton's Philosophy and of the Principal Philosophical Questions Discussed in His Writings.* Boston: W. V. Spencer.

Miner, Valerie, and Helen F. Longino, eds. 1987. *Competition: A Feminist Taboo?* New York: Feminist Press at CUNY.

Mitchell, Sandra D. 2000. Dimensions of Scientific Law. *Philosophy of Science* 67:242–65.

Motterlini, Matteo, ed. 1999. *For and Against Method.* Chicago: University of Chicago Press.

Musgrave, Alan. 1976. Method or Madness. In *Essays in Memory of Imre Lakatos,* edited by Robert Sonné Cohen, Paul K. Feyerabend, and Marx W. Wartofsky. Dordrecht: Reidel.

Newton, Isaac. [1687] 1999. *The Principia: Mathematical Principles of Natural Philosophy.* Translated by I. B. Cohen and A. M. Whitman. Berkeley: University of California Press.

Newton-Smith, William H. 1981. *The Rationality of Science.* Boston: Routledge & Kegan Paul.

Passmore, John. 1966. *A Hundred Years of Philosophy.* 2d ed. New York: Penguin.

Pearl, Judea. 2000. *Causality: Models, Reasoning, and Inference.* Cambridge: Cambridge University Press.

Platt, John R. 1964. Strong Inference. *Science* 146:347–53.

Popper, Karl R. 1935. *Logik der Forschung: Zur Erkenntnistheorie der Modernen Naturwissenschaft*. Vienna: J. Springer.

———. 1959. *The Logic of Scientific Discovery*. New York: Basic Books.

———. 1963. *Conjectures and Refutations: The Growth of Scientific Knowledge*. London: Routledge & Kegan Paul.

———. 1970. Normal Science and Its Dangers. In *Criticism and the Growth of Knowledge*, edited by Imre Lakatos and Alan Musgrave. Cambridge: Cambridge University Press.

Porter, Roy. 1998. *The Greatest Benefit to Mankind: A Medical History of Humanity*. New York: W. W. Norton.

Preston, John, Gonzalo Munévar, and David Lamb, eds. 2000. *The Worst Enemy of Science? Essays in Memory of Paul Feyerabend*. Oxford: Oxford University Press.

Provine, William B. 1971. *The Origins of Theoretical Population Genetics*. Chicago: University of Chicago Press.

Psillos, Stathis. 1999. *Scientific Realism: How Science Tracks Truth*. New York: Routledge.

Putnam, Hilary. 1974. The "Corroboration" of Theories. In *The Philosophy of Karl Popper*, edited by Paul Arthur Schilpp. The Library of Living Philosophers. La Salle, IL: Open Court.

———. 1975. *Mind, Language, and Reality*. Cambridge: Cambridge University Press.

———. 1978. *Meaning and the Moral Sciences*. London: Routledge and Kegan Paul.

Quine, Willard V. 1953. Two Dogmas of Empiricism. In *From a Logical Point of View*, by Willard V. Quine. Cambridge: Harvard University Press. First published, with slight differences, in *Philosophical Review* 60 (1951): 20–43.

———. 1969. Epistemology Naturalized. In *Ontological Relativity and Other Essays*. New York: Columbia University Press.

———. 1990. *Pursuit of Truth*. Cambridge: Harvard University Press.

Railton, Peter. 1981. Probability, Explanation, and Information. *Synthese* 48:231–56.

Ray, Thomas S. 1992. An Approach to the Synthesis of Life. In *Artificial Life II*, edited by Christopher G. Langton, Charles Taylor, J. Doyne Farmer, and Steen Rasmussen. Proceedings of the Workshop on Artificial Life, February 1990, Santa Fe, NM. Redwood City, CA: Addison-Wesley.

Reichenbach, Hans. 1938. *Experience and Prediction: An Analysis of the Foundations and the Structure of Knowledge*. Chicago: University of Chicago Press.

———. 1951. *The Rise of Scientific Philosophy*. Berkeley: University of California Press.

Resnik, Michael D. 1987. *Choices: An Introduction to Decision Theory*. Minneapolis: University of Minnesota Press.

Ricketts, Thomas G. 1982. Rationality, Translation, and Epistemology Naturalized. *Journal of Philosophy* 79:117–35.

Roberts, William A. 1998. *Principles of Animal Cognition*. New York: McGraw-Hill.

Rorty, Richard. 1982. *Consequences of Pragmatism*. Minneapolis: University of Minnesota Press.

Routledge Encyclopedia of Philosophy. 1998. Ed. Edward Craig. London: Routledge.

Salmon, Wesley C. 1981. Rational Prediction. *British Journal for the Philosophy of Science* 32:115–25.

———. 1984. *Scientific Explanation and the Causal Structure of the World.* Princeton, NJ: Princeton University Press.

———. 1989. *Four Decades of Scientific Explanation.* Minneapolis: University of Minnesota Press.

———. 1998. Scientific Explanation: Causation and Unification. In *Causality and Explanation,* by Wesley C. Salmon. Oxford: Oxford University Press.

Schilpp, Paul Arthur, ed. 1974. *The Philosophy of Karl Popper.* The Library of Living Philosophers. La Salle, IL: Open Court.

Schlick, Moritz. 1932–33. Positivism and Realism. *Erkenntnis* 3:1–31 (in German). Translated by Peter Heath and reprinted in *The Philosophy of Science,* edited by Richard Boyd, Philip Gasper, and J. D. Trout (Cambridge, MA: MIT Press, 1991).

Schuster, John A. 1990. The Scientific Revolution. In *Companion to the History of Modern Science,* edited by R. Colby, G. N. Cantor, J. R. R. Christie, and M. J. S. Hodge. London: Routledge.

Shapin, Steven. 1982. History of Science and Its Sociological Reconstructions. *History of Science* 20:157–211.

———. 1994. *A Social History of Truth: Civility and Science in Seventeenth-Century England.* Chicago: University of Chicago Press.

———. 1996. *The Scientific Revolution.* Chicago: University of Chicago Press.

Shapin, Steven, and Simon Schaffer. 1985. *Leviathan and the Air-Pump: Hobbes, Boyle, and the Experimental Life.* Princeton, NJ: Princeton University Press.

Skyrms, Brian. 2000. *Choice and Chance: An Introduction to Inductive Logic.* 3d ed. Belmont, CA: Wadsworth/Thomson Learning.

Smart, J. J. C. 1963. *Philosophy and Scientific Realism.* New York: Humanities Press.

———. 1968. *Between Science and Philosophy.* New York: Random House.

Smith, Adam. [1776] 1976. *An Inquiry into the Nature and Causes of the Wealth of Nations.* Edited by R. H. Campbell and A. S. Skinner. Oxford: Oxford University Press.

Sobel, Dava. 1999. *Galileo's Daughter: A Historical Memoir of Science, Faith, and Love.* New York: Walker.

Sober, Elliott. 1988. *Reconstructing the Past: Parsimony, Evolution, and Inference.* Cambridge, MA: MIT Press.

———. Forthcoming. *Learning from Logical Positivism.* Cambridge: Cambridge University Press.

Sokal, Alan. 1996a. A Physicist Experiments with Cultural Studies. *Lingua Franca* (May–June): 62–64.

———. 1996b. Transgressing the Boundaries: Toward a Transformative Hermeneutics of Quantum Gravity. *Social Text* 14:217–52.

Solomon, Miriam. 2001. *Social Empiricism.* Cambridge, MA: MIT Press.

Sosa, Ernest, and Michael Tooley, eds. 1993. *Causation.* Oxford: Oxford University Press.

Spirtes, Peter, Clark N. Glymour, and Richard Scheines. 1993. *Causation, Prediction, and Search*. New York: Springer-Verlag.

Stalker, Douglas Frank. 1994. *Grue! The New Riddle of Induction*. Chicago: Open Court.

Stanford, P. Kyle. 2000. An Antirealist Explanation of the Success of Science. *Philosophy of Science* 67:266–84.

———. 2001. Refusing the Devil's Bargain: What Kind of Underdetermination Should We Take Seriously? *Philosophy of Science* 68:S1–S12.

Sterelny, Kim. 1994. Science and Selection. *Biology and Philosophy* 9:45–62.

Stich, Stephen P. 1983. *From Folk Psychology to Cognitive Science: The Case against Belief*. Cambridge: MIT Press.

Strevens, Michael. 2003. The Role of the Priority Rule in Science. *Journal of Philosophy* 100:55–79.

Sulloway, Frank J. 1996. *Born to Rebel: Birth Order, Family Dynamics, and Creative Lives*. New York: Pantheon Books.

Suppe, Frederick, ed. 1977. *The Structure of Scientific Theories*. Chicago: University of Illinois Press.

Suppes, Patrick. 1981. The Plurality of Science. In *PSA 1978*, vol. 2, edited by Peter D. Asquith and Ian Hacking, 3–16. East Lansing, MI: Philosophy of Science Association.

———. 1984. *Probabilistic Metaphysics*. Oxford: Blackwell.

Toulmin, Stephen Edelston. 1972. *Human Understanding*. Princeton, NJ: Princeton University Press.

Toulmin, Stephen Edelston, and June Goodfield. 1962. *The Fabric of the Heavens: The Development of Astronomy and Dynamics*. New York: Harper.

———. 1982. *The Architecture of Matter*. Chicago: University of Chicago Press.

van Fraassen, Bas C. 1980. *The Scientific Image*. Oxford: Oxford University Press.

Wason, Peter Cathcart, and Philip N. Johnson-Laird. 1972. *Psychology of Reasoning: Structure and Content*. Cambridge, MA: Harvard University Press.

Westfall, Richard S. 1980. *Never at Rest: A Biography of Isaac Newton*. Cambridge: Cambridge University Press.

———. 1993. *The Life of Isaac Newton*. Cambridge: Cambridge University Press.

Wittgenstein, Ludwig. [1922] 1988. *Tractatus Logico-Philosophicus*. Translated by David Pears and Brian McGuinness. London: Routledge and Kegan Paul.

———. 1953. *Philosophical Investigations*. Translated by G. E. M. Anscombe. New York: Macmillan.

Wolfe, Tom. 1998. *A Man in Full: A Novel*. New York: Farrar Straus & Giroux.

Woolgar, Steven. 1988. *Science: The Very Idea*. London: Ellis Horwood.

Worrall, John. 1989. Structural Realism: The Best of Both Worlds? *Dialectica* 43:99–124.

Ziman, John M. 2000. *Real Science: What It Is and What It Means*. Cambridge: Cambridge University Press.

Index